Thinking Like a Mountain

Thinking Like a Mountain

ALDO LEOPOLD AND THE EVOLUTION
OF AN ECOLOGICAL ATTITUDE
TOWARD DEER, WOLVES, AND FORESTS

SUSAN L. FLADER

UNIVERSITY OF MISSOURI PRESS

1974

Permission has been granted by the University of Wisconsin Archives for the reproduction of materials on pages x and xi; by the University of Wisconsin Department of Wildlife Ecology for the materials on pages xiii and xiv; by the Forest Service for the photographs on pages ix and xii; and by Robert A. McCabe for the material on pages xv and xvi.

The excerpts from the manuscript version of "Thinking Like a Mountain" were first published in *A Sand County Almanac,* copyright © 1949, Oxford University Press.

Permission has been granted by the University of Wisconsin Archives for reproduction of the graph on p. 73 and by the Wisconsin Department of Natural Resources for the graphs on p. 203.

For Dad

Contents

Aldo Leopold (second from right) with a reconnaissance crew of the U.S. Forest Service. Leopold's first job after he graduated from Yale Forest School in 1909 was to lead this crew to map and cruise timber in the Apache National Forest in Arizona Territory.

THE PINE CONE

JANUARY, 1919 (12th ISSUE)
1500 MEMBERS
OFFICIAL BULLETIN OF THE NEW MEXICO GAME PROTECTIVE ASSOCIATION
ISSUED QUARTERLY
CIRCULATION, 5000

OUR PLATFORM

1. We stand for rigorous and impartial enforcement of the game and fish laws.
2. We stand for federal control of migratory birds and prohibition of spring shooting.
3. We stand for co-operation with stockmen in a vigorous campaign against predatory animals.
4. We stand for an adequate system of Game Refuges.
5. We stand for such an increase in game and fish as will furnish legitimate sport for every citizen.
6. We are opposed in general to the public propagation in New Mexico of foreign species as a substitute for native American game.
7. We represent 1,500 members, each and every one pledged to observe the letter of the law and the spirit of good sportsmanship.
8. We are not in politics.
9. We stand behind every warden who does his duty.
10. We offer $50.00 reward for information leading to the arrest and conviction of any person killing antelope, mountainsheep or ptarmigan.

"The ways of outdoor life, the nobility of courage, the joy of beauty, the blessedness of enough, the glory of service, the power of kindness, the superexcellence of peace of mind, and the scorn of death —these were the things the Redman stood for. These were the sum of his faith."

ERNEST THOMPSON SETON.

As the vine scatters the seeds of the pine and fir tree, so may this little paper scatter the seeds of wisdom and understanding among men.

SIX RULES FOR SPORTSMEN

1. *Be a Real Sportsman.* There is more honor in giving the game a square deal than in getting the limit.
2. *Make Sure It's a Buck.* If you can't see his horns—she hasn't got any.
3. *Help Enforce the Game Law.* Game and fish are public property and only a game-hog will take more than his fair and legal share. Violations should be reported to the nearest Deputy Warden, Forest Ranger, or Game Protective Association.
4. *Respect the Ranchman's Property.* He regards the man who leaves his gates open, cuts his fences, chouses his livestock, or shoots near dwellings, as an outlaw. Put yourself in his place.
5. *Be careful With Your Campfire and Matches.* One tree will make a million matches; one match can burn a million trees.
6. *Leave a Clean Camp and a Clean Record.* Unburied garbage, crippled game, and broken laws, are poor monuments for a sportsman to leave behind him.

Why Governor Larrazolo Should Re-appoint Rouault

The job of re-building New Mexico game resources is just at this time in a critical stage.

For two years previous to the war, the G. P. A. was engaged in a campaign of public education. This was a necessary foundation to successful future action.

During the war, the whole game protection movement necessarily had to "mark time."

Now that the war is over, the time has come to go after results,—to undertake actual constructive work which will produce more game.

It stands to reason that the state game department is the most essential single part of the machinery by which more game is to be produced.

Therefore the lack of an efficient and sympathetic state game department would block the whole program of game conservation in this state.

There is no use mincing words over the fact that Mr. Rouault is the first New Mexico Game Warden who has come any where near giving our sportsmen efficient and sympathetic service. Therefore the G. P. A., representing the organized sportsmen of the state, want him reappointed, and considers it only reasonable to expect that Governor Larrazolo will act accordingly.

There are two essential points about our insistence on the reappointment of Mr. Rouault which should be thoroughly understood.

First, we are after a man for the job, not a job for the man. There may be other candidates for the appointment who believe they are better qualified than Mr. Rouault. If so, let them place their qualifications before the sportsmen of the state, instead of trying to get in by the back door. The fact that all other candidates have so far tried the back door route only speaks for itself.

Second, we are asking the Governor not for a favor, but for a right. The sportsmen support the state game department. The sportsmen are principally affected by its efficiency or inefficiency. The stockmen are never saddled with a sanitary board unsatisfactory to them. Likewise, the organized sportsmen should not be saddled with a Game Warden whom they do not approve.

Fortunately, Governor Larrazolo has given us every reason to believe that he appreciates the validity of the forgoing arguments. We confidently expect that he will act accordingly.

Planting fish fry in New Mexico streams and lakes is an investment that will pay high dividends in sport, health, and food.

VARMINTS

GAME PROTECTION in New Mexico is going to *make or break* on predatory animals.

Good game laws well enforced will raise enough game either for sportsmen or for varmints, but not enough for both.

Either the lions, or the game, must go.

NEW MEXICO is *leading the West* in the campaign for eradication of predatory animals.

Our State Council of Defense has pooled its dollars with the U. S. Biological Survey in a mighty effort to rid the ranges of these pests. This effort has hung varmint scalps on the clothline, but it has just begun.

It must keep on.

AS long as the Council of Defense stays on the job, it *will* keep on. But how long will this be!

It is for the Legislature to decide.

The point is that the sportsmen of New Mexico *hold the Legislature responsible* for continuing the predatory animal work.

THIS means dollars, not in dribbles, but in five figures,—dollars without stint or limit.

Quick work on varmints is the cheapest. Slow work, and bounties, merely remove the increase, and are sheer waste of money.

The sportsmen and the stockmen—one third the population and one-half the wealth of New Mexico—demand the eradication of lions, wolves, coyotes, and bobcats.

DEPREDATIONS of varmints do not make game protection useless. On the contrary game protection makes the killing of varmints necessary.

The fact that a weasel is getting away with your pullets is no good reason for wringing the necks of your hens.

It is, however, the very best reason for going out and killing the weasel.

The fact that lions are getting away with our deer is no good reason for killing off what few deer we have left.

It is most emphatically a reason for going out after the last lion scalp, and getting it.

THE SPORTSMEN *WANT ACTION* ON VARMINTS.

BEWARE OF UNCLE SAM!

PUTS THE FEDERAL LID ON POACHING IN THE NATIONAL FORESTS. HUNTERS MUST OBEY THE STATE LAW.

The Forest Service has always looked with favor upon the protection of game. It has, from the start, regarded game as a forest resource which, although according to legal opinion belonging to the State rather than to the nation, is worth preserving and handing down, along with the timber and water and forage, unimpaired to posterity. It has given of its time and money to put this idea into practice. Its officers have landed and prosecuted many a game-hog and many a poacher.

Now comes a radical step forward.

The Secretary of Agriculture, head of the Department of which the Forest Service is a Bureau, has promulgated a regulation that forbids the going or being upon a National Forest with intent to catch, trap, wilfully disturb or kill any kind of game animal, game or non-game bird or fish, or take the eggs of any such bird, contrary to the provisions of the State law. This regulation is based upon full authority delegated by Congress to the Secretary of Agriculture. This regulation has the force of law. It means:

(1) That violators of the State laws within a National Forest are subject to federal prosecution. As these Forests comprise practically all the big-game hunting grounds of New Mexico, the sweeping force of the regulation is obvious.

(2) That the regulation is so framed as to require only proof of INTENT on the part of the law violator.

(3) That Uncle Sam has seen and recognized the danger of game extermination, and with all his might, is going to help the States enforce the law.

SUPERSTITIONS; OR PASSING THE BUCK

Ask a mountaineer, a rancher, a sheepherder, or any other dweller of the open: "Who is responsible for the extermination of game!", and he will reply without the flicker of an eyelash:

"THE CITY HUNTER."

Ask the city hunter—from the urchin with a 22-short to the de luxe sport with a six-cylinder pump-gun and a twin-six roadster—and he will reply without a single twinge of the conscience:

"THE MOUNTAINEER, THE RANCHER, THE SHEEPHERDER."

Get two specimens of these two species of the genus homo together and propound the same riddle. With singular unanimity and without a single blush of shame, they will both bellow:

"THE INDIAN."

Individually, they are all wrong, but collectively they are absolutely right; for it is the city hunter plus the mountaineer, plus the rancher, plus the sheepherder, plus the Indian, plus everybody else but me, I, myself, that is exterminating the game.

Incidentally the game of New Mexico held its ground and flourished for Heaven knows how many centuries or scores centuries against the combined onslaughts of the Indians and the beasts of prey. It was not until there appeared upon the scene HIS MAJESTY THE WHITE MAN that the real devastation began.

Let's forget these infantile superstitions and pull together to save the remnants.

There is an entente cordiale between the beaver and the farmer of the Rio Grande Valley. Both are put to it to find enough wood, the one for food, the other for fuel. Bre'r Beaver with his extraordinary teeth, fells the cottonwoods along the river fringe of bosques, cuts off the twigs and branches, gnaws them into short lengths, and stores them in caves beneath the bank, below-the water line. This gives him a supply of juicy and apparently palatable bark for his winter's food. Mr. Farmer hauls away the trunk of the tree for firewood. Nothing is left but the stump and the chips. The cottonwood groves do not seem to suffer any real damage from the logging operations of the beaver.

Shooting ducks by moonlight is not only illegal, but a foolish want of ducks and ammunition.

To the long list of extinct animals, it is to be hoped that the game-hog will soon be added.

The Pine Cone (above), a newspaper edited by Leopold. Through it he promoted a program to increase game populations and eradicate "varmints." A letter from former President Theodore Roosevelt (top right) commended these efforts. Leopold's wife and a buck Leopold shot (bottom right) on a 1923 pack-saddle trip in the Gila National Forest, where his game program had been implemented and where he was promoting a wilderness area.

Metropolitan

THE LIVEST MAGAZINE IN AMERICA
432 FOURTH AVENUE, NEW YORK

Office of
Theodore Roosevelt

January 16, 1917.

My dear Mr. Leopold:

      Through you, I wish to congratulate the Albuquerque
Game Protective Association on what it is doing.   I have just
read the Pine Cone.   I think your platform simply capital, and
I earnestly hope that you will get the right type of game warden.
It seems to me that through your association, New Mexico is setting an
example to the whole country.

          Sincerely yours,

          Theodore Roosevelt

Mr. Aldo Leopold, Secretary,
Albuquerque Game Protective Ass'n,
Albuquerque, N. Mex.

Evidence of severe overbrowsing by deer in Black Canyon (above), within the Gila Wilderness. Leopold received reports of damage there a few years after he left the Southwest in 1924. A construction camp along the North Star Road (below). The road bisected the wilderness to give hunters access to the deer.

A slick-and-clean spruce forest in Germany (above). Leopold was appalled at the artificialized management he observed there in 1935. Along the Río Gavilán in Mexico (below) the following year, he first observed a forest that had maintained biotic integrity.

A pile of starved deer in Wisconsin (above). The conservation department photographed these deer to dramatize the problems of an overpopulated deer range. In *Save Wisconsin's Deer* (below), irate citizens attacked Leopold's continued advocacy of herd reduction and his defense of predators after the "crime of '43," when many does and fawns were killed.

Prof. Aldo Leopold
424 University Farm Place
Madison, Wis.

PUBLISHED IN THE INTERESTS OF THE WILDLIFE AND NATURAL RESOURCES OF WISCONSIN

OFFICIAL PUBLICATION OF THE "SAVE WISCONSIN'S DEER" COMMITTEE OF WISCONSIN

SAVE WISCONSIN'S DEER

ISSUED MONTHLY   JUNE, 1945   NUMBER ELEVEN

'Assume They Know More Than I Do'--Leopold

Bambi Says . . .

A suggestion to the commission: Why not cross a carrier pigeon with an ordinary woodpecker? The idea is to get a bird that will not only deliver a message, but also knock on the door!

Then, when the experts make a survey on a bird like this, all they need do is ask how many people heard knocking on their doors. This survey should eliminate all guessing.

This idea is as good as the one where the fellow tried to teach glow worms to give people a "hot foot!"

There's an old saying that it's a woman's privilege to change her mind, but now the commission wants to follow suit. They used to tell us that feeding deer was "impractical," and the boys up north showed them it could be done. Now they try to tell us it's too expensive! Who pays for it?—the sportsmen! Nope, the commission is not in the red—they're nursing a pretty surplus!

The sportsmen gladly paid to give our deer a little extra "handout" during the winter and if we have anything to say—it's going to be spent for the purpose intended!

Things like this remind us of

CON. BULLETIN SHOULD CONTINUE VIOLATION LIST

Many persons feel that the discontinuance of the publishing of names of persons convicted of fish and game violations, by the Conservation Bulletin is a grave error.

The publishing of this list has not only prevented many people from violating, but it also gives everyone the satisfaction of knowing just who they are. It has also prevented many from "repeating" rather than face the publicity following a conviction.

It may be possible that certain people favor the Bulletin discontinuing this practice, but to those who infringe on the game laws it is a blessing.

Deer Kill Figures To Be Released

It has been stated that the figures on the 1944 deer kill will be released and the public can brace themselves for one of the greatest "fairy tales" ever told from Madison.

We are advised that the kill is placed at 35,345 (based on monthly estimates) for the entire deer counties. Bayfield is listed with a kill of 2,757, Vilas county 2,408, Jackson county 1,944, Oneida county 1,843, and Marinette county 1,696 deer.

This would approximately average up as one hunter out of every three and one-half succeeding in getting a deer. It is based on a 32 per cent return on game cards for 120,147 tags sold. Our survey

Public Hearing On Control Bill Proves Lively

The public hearing on Bill No. 581 A, known as the "Controlled Shooting Bill", was held by the joint legislatives committees at Wausau on May 21st, substantiated the old saying by General Sherman, "War is hell". The proponents, as well as the opponents, gave no quarter and asked no quarter. There was never a dull moment.

Assemblyman John Young (Oconto) acted as chairman for the senate and assembly conservation committees. They came north to get a true cross section of the people's attitude on the bill which would grant the commission considerable more power. "It shall not pass" was the battle-cry of the opponents of the bill.

The Joint committees consisted of the following: Senators Wm. McNeagh, 25th; Phillip Downing, 30th; Ernest Heden, 12th. Senators Taylor Brown, 19th (chairman) and Robert Tehan, 9th, were unable to attend due to other official business.

See PUBLIC HEARING—Page 2

IRON COUNTY CLUB GAINING MEMBERS

Although the Iron County Sportsmen club was organized but a few months ago, it is making wonderful progress. With a membership today of approximately 275, the energetic officers

CIRCUIT COURTS MAY SETTLE DEER FIGHTS

It has been stated that a number of sportsmen in Vilas county as well as in several other counties, will petition the circuit courts of their respective counties, in the event that a deer season is imposed by the conservation commission.

In the past it has been the practice of the commission to totally disregard the wishes of certain counties, even though such counties have requested closed deer seasons, through their county game meetings. It is felt that several applications for injunctions would improve the situation which now exists and determine if the people have any recourse whatsoever.

It is also felt that if the present plans of several sportsmen's groups are carried out it will encourage other counties to dissatisfied with certain conservation orders, to take like action in the future

Check on Lamprey Damage Control

Madison, Wis.—The possibility of working out methods to control the marine lamprey, an eel-like creature that preys on fish in the Great Lakes, is under investigation by the conservation department.

Lampreys attach themselves to fish with their suction mouth parts and feed on fish blood. The fish are scarred even when they

Coyote Stand Also Contradicts Game Survey

(By The "OBSERVER")

"Those who assume that we would be better off without any wolves are assuming more knowledge of how nature works than I claim to possess."

This statement comes from none other than Professor Aldo Leopold, one of Wisconsin's conservation commissioners, and appeared in the April issue of the Conservation Bulletin. Read it again because it has that touch of "Leopoldian egotism" and insinuates that he, the great Aldo, places his knowledge above that of any Wisconsin citizen. For the benefit of our readers who do not receive the Bulletin, we publish excerpts of the article with our comments.

He Thinks' Herd Is Smaller

"The past deer hunting season shows that the deer herd has shrunk more in one year than I expected it to. Among the causes of this shrinkage are the 1943 doe season and the 1942 starvation of fawns. I doubt if the herd has shrunk too much in the average county."

Well, at last (after publicly admitting his guess on the deer herd was too high) he now admits that the herd was reduced far more than he thought.

His 'Fundamental' Principle

"I now favor restoring the

Leopold admiring a pine, one of thousands he and his family planted at their sand-country farm in central Wisconsin in their effort to restore the land to ecological integrity.

The original pencil draft of "Thinking Like a Mountain,"
1 April 1944, showing numerous refinements, including
erasures and cutting and pasting.

Preface

Aldo Leopold (1887–1948) is best known as the author of *A Sand County Almanac* (1949), a volume of nature sketches and philosophical essays recognized as one of the enduring expressions of an ecological attitude toward man and land. To many who know him through these essays, he is akin to Thoreau because of his keen observation, his philosophic penetration, and his clarity of expression. Yet he was also an internationally respected scientist and conservationist, instrumental in formulating policy and building ecological foundations for two new professions in twentieth century America, forestry and wildlife management. During his professional career he published several books and nearly 350 articles, most of them on scientific or policy matters.

His life spanned a period during which scientific and technical advances were being made that could provide support for a fundamental change in the relationship of Americans to their environment. Leopold as scientist was acquainted with the leading ecologists of his day; as forest administrator, professor, and consultant to government agencies and private organizations, he had the ear of policymakers and access to land on which his ideas could be tried. His concern was not only with advancing ecological science, but with the communication of new findings and the development of techniques and incentives for applying them at the level of the land manager, whether forester, farmer, or small landowner like himself.

Through all his efforts he was dedicated to the conviction that we would never solve our conservation problems on a large scale until we as a people had attained an ecological attitude toward our environment. This attitude would be the basis of an ethic for the use of the land. The notion of a land ethic was rooted in his perception of the environment, and that perception was deepened and clarified throughout his life by his own observations and experiences and by new findings in ecological science. Although the notion is coming into vogue today, many who fancy themselves disciples of Leopold shun the rigorous thinking and constructive openmindedness that to him were essential.

This study analyzes the development of Leopold's thought at a level where observation and experience, science and philosophy, policy and politics converge on a single problem running through time. The introductory chapter presents the intellectual problem of deer–wolf–forest interrelationships as symbolized in Leopold's essay, "Thinking Like a Mountain," then offers a biographical sketch of his professional career and a brief analysis of the development of his philosophy, placed in the context of historical developments in ecology.

The next two chapters are concerned with his experiences in the national forests of the Southwest. Chapter 2 traces the development of his ideas on environmental change and game management up to 1927, when he wrote a book-length manuscript, "Southwestern Game Fields." Chapter 3, "The Gila Experience," although it pertains mostly to the years 1927–1931, when Leopold was already in Wisconsin, shows how his ideas were modified by a deer "irruption" that forced the first major dismemberment of the Gila Wilderness in New Mexico. The following chapter shifts to the Midwest and other regions; it discusses Leopold's increasing emphasis on environmental management, research, and public policy during the 1930s and the philosophical impact of his experiences in Germany and in northern Mexico at mid-decade. Taken together, chapters 2, 3, and 4 deal with the development of Leopold's ideas about the techniques, or means, of wildlife management and the broadening of his perspective on the objectives, or ends. They analyze his transition from an emphasis on resource supply and environmental "control" to an ecological concern with land health and a philosophy of naturally self-regulating systems.

In the last two chapters attention shifts from the evolution of Leopold's own ecological attitude to his efforts to stimulate the development of similar attitudes and values in the public mind. Chapter 5, "Too Many Deer," deals with the deer irruption in Wisconsin in the early 1940s and Leopold's first efforts, up to about 1944, to win support for or at least acquiescence in necessary herd reduction. "Adventures of a Conservation Commissioner" carries the public opinion and policy issues up to his death in 1948 and considers Leopold's analysis of some of the ecological issues raised by the deer problem in the 1940s. It examines his conception of the responsibility of the scientist and attempts to pose some questions as to ways in which his activity in the public arena in a crisis situation may have impeded his ecological analysis even as it strengthened his conviction of the need for a land ethic to deal with complex, highly dynamic environmental problems.

Aldo Leopold's changing attitude toward deer and deer–wolf–forest interrelationships is but one current in his intellectual experience, no more important than a number of others that might be identified. It is a fairly pervasive one, however, and can be followed through almost every phase of his career as it mingles with other currents in his developing thought. He was involved with deer, wolves, and forests as hunter and as game protectionist, as forest officer and as wildlife manager, as scientist, consultant, and educator, as citizen and as public policymaker, as environmental philosopher, and even as propagandist. These roles influenced in subtle ways his understanding and interpretation of events and thus contributed to the working out of his philosophy.

It may seem strange to study such a subject as attitudes toward deer, wolves, and forests for keys to a person's philosophical development. For me, however, this study has provided a way of analyzing the progression in Aldo Leopold's thinking, because he wrote so much on the subject, at so many different times of his life, and in so many different capacities. Moreover, his thinking and writing about deer and deer–wolf–forest interrelationships were repeatedly tested by the imperative of action and the lessons of experience, and time and again he was compelled to alter his views on particular issues. Leopold always had a reasoned, philosophical basis for his writing and his action, but his fundamental values remained nominally the same throughout his life; they were developed by clarification more than by outright change. Consequently, it can be difficult to see a progression in his thought unless one focuses on his thinking about a particular subject in all its concrete detail. Through attention to details, it is possible to follow the changing pattern of his ideas, to see how he incorporated new observations into his image of ecological interrelationships, adjusting the image incrementally with each new observation or bit of information until, sometimes, a certain experience triggered a rearrangement of elements, resulting in a new interpretation. It was through such a process—through observation, experience, and reflection—that Aldo Leopold evolved his mature philosophy.

This study deals not alone with the development of Leopold's thinking but also with his efforts to promote action toward solution of the problem of overbrowsing by deer, which in turn involved him in attempts to develop ecological attitudes at various levels in the political realm. Deer problems provide an apt focus for studying the evolution of public attitudes and policies because they were very much a public issue. Not only were deer of social and economic value as shootable big game and as visible wildlife, but they also impinged on other matters of actual or potential public concern, such as vegetational diversity, forest reproduction, logging practices, livestock grazing, soil erosion, predator control, the fortunes of other wildlife species—or, in the largest sense, the health of the land.

The nature of deer problems and their impact on other public concerns changed radically during the course of Leopold's life. His intellectual equipment for understanding the problems changed also, following upon conceptual developments in ecological science and research findings in game management. The nature of the problems, however, changed almost faster than his capacity for understanding them. On occasion he did not recognize emerging problems until they had reached near-crisis proportions. And at that point the public, at most levels where action might be taken toward a solution, was still in blissful ignorance—or worse, zealously attacking a problem that no longer existed.

We are concerned, therefore, not with the unfolding of one individual's understanding of a static situation but

Preface

with the evolution of an ecological attitude toward a dynamic situation, by an individual and by the public, not separately but as they interacted. This study is not intended to be a case history in either applied ecology or public policy, though it may provide some insights in both areas. Rather, it is intended as a study in the evolution of a way of thinking, at both personal and public levels. It seeks to probe what it means to think ecologically and to deal with an ecological issue in the public arena.

S.L.F.
Columbia, Missouri
June 1974

Acknowledgments

In my work on this study, I am most indebted to the Leopold family for their total cooperation, penetrating insights, and encouragement, given during hours of conversation and days afield. Members of the immediate family are Mrs. Aldo Leopold of Madison; A. Starker Leopold, Professor of Wildlife Ecology, and Luna B. Leopold, Professor of Geology, both at the University of California, Berkeley; Nina Leopold Bradley of Bozeman, Montana; A. Carl Leopold of the Department of Horticulture at Purdue University, Lafayette, Indiana; and Estella B. Leopold, research paleobotanist with the U.S. Geological Survey, Denver. Leopold's brother Frederic Leopold and his sister Marie Leopold Lord of Burlington, Iowa, also gave freely of their recollections.

In its conception this study benefited immeasurably from the wisdom and generosity of the late David M. Potter of the Department of History at Stanford University. Others who generously offered help and advice during conceptualization and writing include Wallace Stegner, Professor Emeritus of English, and Don E. Fehrenbacher, Professor of History, both of Stanford; Merle Curti, Professor Emeritus of History, and Harold C. Jordahl, Jr. of the Department of Urban and Regional Planning at the University of Wisconsin, Madison; and Irving K. Fox, Director of the Westwater Research Centre at the University of British Columbia, Vancouver.

I am grateful to many individuals for insights on various phases of Leopold's career and problems with which he was involved. Arthur Ringland, the late Raymond Marsh, and the late L. F. Kneipp, of Washington, D.C., and Elliott Barker of Santa Fe, all colleagues of Leopold's in the U.S. Forest Service, reminisced about early days in the Southwest, while Samuel Servis of Silver City, New Mexico, E. A. Tucker of Albuquerque, and D. I. Rasmussen of Ogden, Utah, provided additional historical information. Leopold's early contributions in the field of game management were recalled by Herbert L. Stoddard of Thomasville, Georgia, Wallace B. Grange of Calio, North Dakota, and W. Noble Clark and the late A. W. Schorger of Madison; W. J. Harris, Jr. of Grosse Pointe Park, Michigan, discussed his work for the Huron Mountain Club. Many participants in Wisconsin's deer-forest controversy offered their observations, among them William Feeney and the late Ernest Swift, formerly of the Wisconsin Conservation Department; the late Charles F. Smith and W. J. P. Aberg of the Conservation Commission; Clarence Searles and H. O. Schneiders of the Conservation Congress; Otis Bersing, Burton Dahlberg, James B. Hale, Cyril Kabat, and R. E. Wendt of the Department of Natural Resources; and Forest Stearns of the University of Wisconsin, Milwaukee.

Joseph J. Hickey and Robert McCabe, students of Aldo Leopold and now professors in the University of Wisconsin Department of Wildlife Ecology, answered innumerable questions and helped in many ways. Other students and colleagues of Leopold who were particularly helpful to me in connection with this study include Clay Schoenfeld and Arthur Hasler of the University of Wisconsin, Frederick and Frances Hamerstrom of Plainfield, Wisconsin, Lyle K. Sowls of the University of Arizona, Tucson, H. A. Hochbaum of Delta, Manitoba, and William H. Elder of the University of Missouri, Columbia. Many other students and friends of Aldo Leopold have shared their memories with me, and I thank them.

My research for this study was greatly facilitated by J. E. Boell and J. Frank Cook of the University of Wisconsin Archives; E. B. Fred, President Emeritus of the University of Wisconsin; Patricia M. Schleicher of the University of Wisconsin Department of Wildlife Ecology; and Walter E. Scott of the Wisconsin Department of Natural Resources. I have also received helpful assistance from members of the staffs of the National Archives, the Federal Records Center at Denver, the Conservation Library Center of the Denver Public Library, the U.S. Forest Products Laboratory in Madison, Region 3 of the U.S. Forest Service, the Gila National Forest, the New Mexico Department of Fish and Game, the Wisconsin Department of Natural Resources, the State Historical Society of Wisconsin, the Yale Forest School, the Forest History Society, and other libraries,

Acknowledgments

agencies, and institutions. This study and research for the larger biography of which it is a part were supported by fellowships and research grants from Stanford University, Resources for the Future, Inc., the University of Wisconsin, and the University of Missouri.

Many colleagues, students, and friends have read drafts of the manuscript and offered suggestions. I should like to thank all of them, and especially Genevieve Bancroft, Kenneth Bowling, Robert Cook, William Cronon, Dolores Flader, James Hale, Frances Hamerstrom, James Harris, Ruth Hine, Daniel Kozlovsky, Kathryn Larson, A. Starker Leopold, Clay Schoenfeld, Lyle Sowls, and Curtis Synhorst.

Finally, may I express my appreciation for the title-page drawing contributed by Charles W. Schwartz.

To all these people and many others I am deeply grateful. The greatest reward of a study like this is the friends one acquires in the process.

1

Thinking Like a Mountain

On the first day of April 1944, Aldo Leopold sat down with sharpened pencils and a pad of yellow, blue-lined paper, prepared to acknowledge that he had once felt very differently about what he now regarded as the essence of an ecological attitude.

"A deep chesty bawl echoes from rimrock to rimrock as it rolls down the mountain and fades into the far blackness of the night," he began. "It is an outburst of wild defiant sorrow, and of contempt for all the adversities of the world." The deer, the coyote, the cowman, the hunter, in each the call instilled some immediate, personal fear or hope: "Only the mountain has lived long enough to listen objectively to the howl of a wolf." Leopold dated his own conviction that there was a deeper meaning in that howl from the day during his years in the Southwest when he had shot a wolf and watched it die:

> We reached the old wolf in time to watch a fierce green
> fire dying in her eyes. I realized then, and have known
> ever since, that there was something new to me in those
> eyes—something known only to her and to the mountain.
> I was young then, and full of trigger-itch; I thought that
> because fewer wolves meant more deer, that no wolves
> would mean hunters' paradise. But after seeing the green
> fire die, I sensed that neither the wolf nor the mountain
> agreed with such a view.[1]

1. "Thinking Like a Mountain," holograph, 1 April 1944, 3 pp., General Files—Aldo Leopold, Series 9/25/10–6 Box 18, University of Wisconsin Division of Archives (hereafter cited as LP 6B18 [Leopold Papers, Series 6, Box 18]; for a guide to abbreviations and a brief description of the various manuscript collections used in this study, see the Bibliographical Note). Leopold used the expression "thinking like a mountain" to characterize objective or ecological thinking; it should not be viewed as personification.

Thinking Like a Mountain

In this essay entitled "Thinking Like a Mountain," Aldo Leopold compressed into one dramatic moment a realization that had required years. It was a realization that grew, as he went on to suggest, out of his lifelong experience with the management of deer on wolfless range:

> Since then, I have lived to see state after state extirpate its wolves. I have watched the face of many a newly wolfless mountain, and seen the south-facing slopes wrinkle with a maze of new deer trails. I have seen every edible bush and seedling browsed, first to anaemic desuetude, and then to death. I have seen every edible tree defoliated to the height of a saddlehorn. . . . In the end the starved bones of the hoped-for deer herd, dead of its own too-much, bleach with the bones of dead sage, or molder under the high-lined junipers.

A buck taken by wolves, he concluded, could be replaced in two or three years, but a range browsed out by an over-population of deer "may fail of replacement in as many decades."

The wolf, as one of the large carnivores, belonged at the very apex of the biotic pyramid, the image employed in ecology to represent the energy circuit of nature. Through millennia of evolution the pyramid had increased in height and complexity, and in Leopold's thinking this elaboration and diversification contributed to the smooth functioning, or health, of the system. Man with his arrogance and his engines of violence now presumed, in his solicitude for deer and cattle, to lop the large carnivores from the apex of the pyramid, making food chains shorter and less complex and thus disorganizing the system. Because the wolf stood at the apex of the pyramid, it became Leopold's symbol of the pyramid itself, of land health. He did not elaborate on this symbolism in "Thinking Like a Mountain," but it is there. It is the hidden meaning in the howl of the wolf. One who could listen objectively to that howl —who could visualize the wolf in its relation to the total life process of the ecosystem through time, not just as it might affect one's own immediate interests—was thinking ecologically, like a mountain.

During his early years in the national forests of the Southwest, Leopold's reaction to the howl had been far from objective. He had been a leader in a campaign by sportsmen and stockmen to eradicate wolves, mountain lions, and other large predators from the deer and cattle ranges of Arizona and New Mexico. "It is going to take patience and money to catch the last wolf or lion in New Mexico," he had told delegates to the National Game Conference in New York in 1920. "But the last one must be caught before the job can be called fully successful."[2]

It was deer that had had a special place in Leopold's affections in those days, and he had written of them as the 'numenon', or inner meaning, of the mountains:

> To the deer hunter or the outdoorsman, deer are the numenon of the Southwestern mountains. Their presence or absence does not affect the outward appearance of the mountain country, but does mightily affect our reaction toward it. Without deer tracks in the trail and the potential presence of deer at each new dip and bend of the hillside the Southwest would be, to the outdoorsman, an empty shell, a spiritual vacuum.[3]

By 1944, when he wrote "Thinking Like a Mountain," the destructive potential of too many deer was all too apparent, and the wolf had taken the place of the deer in Leopold's sentiment as a symbol of ecological integrity.

"Thinking Like a Mountain" was written as one of a series of essays that Leopold intended to publish in a book to illustrate the process of ecological perception and to follow out some of its meanings and implications. It was written at a time when he despaired of having his ideas on deer and wolves accepted by the public or his book of essays accepted by a publisher; and it was written at least in part in response to the urging of a former student of his, H. Albert Hochbaum.

Hochbaum, director of the Delta Duck Station in Mani-

2. "The Game Situation in the Southwest," *Bulletin of the American Game Protective Association*, 9:2 (April 1920), 5.

3. "Southwestern Game Fields," typescript, c. 1927, Ch. III, p. 1, LP 6B10.

toba, who at the time was preparing pen-and-ink drawings for the proposed book, believed that the essay as a whole breathed too deeply of regret and of aloof sourness or self-righteousness toward man's despoliation of nature. "If we always regret what we have done," he wrote Leopold, "we must regret that we are men. It is only by accepting ourselves for what we are, the best of us and the worst of us, that we can hold any hope for the future." What had always impressed him in his personal contacts with Leopold, Hochbaum noted, was Leopold's unbounded enthusiasm for the future and his common sense way of thinking, "not that of an inspired genius, but that of any other ordinary fellow trying to put two and two together." He urged Leopold to acknowledge somewhere in his writings that he had not always felt the way he did now: "Because you have added up your sums better than most of us, it is important that you let fall a hint that in the process of reaching the end result of your thinking you have sometimes followed trails like anyone else that led you up the wrong alleys." In particular he pointed to Leopold's role in planning the extermination of wolves in New Mexico.[4]

The writing of "Thinking Like a Mountain" was thus a milestone for Leopold, and he realized its significance. He immediately sent a copy to Hochbaum and included it with a dozen essays he was sending around to potential publishers. He had once thought to call his book "Marshland Elegy—And Other Essays," he explained to publishers, but "it now strikes me that 'Thinking Like a Mountain' might be a better key to its contents."[5] The book was finally published five years later with a different title, *A Sand County Almanac,* a different illustrator, Charles W. Schwartz of Missouri, and a greatly augmented selection of essays. But "Thinking Like a Mountain" remains the most graphic piece in it and the only one in which Leopold acknowledges a major change in his thinking over the years.

4. H. A. Hochbaum to Aldo Leopold (hereafter cited as AL), 4 Feb. 1944, 11 March 1944, 22 Jan. 1944, LP 6B5.
5. AL to Clinton Simpson, 8 June 1944, LP 6B5.

Evolution and Ecology

Aldo Leopold's intellectual development mirrors the history of ecological and evolutionary thought, while his professional career spanned the first half century of the movement for conservation and resource management in America. His enduring achievement was to integrate the two strands—the scientific basis and the conservation imperative—in a compelling ethic for our time.

Ecological science had its roots in the evolutionary thought of Charles Darwin. The term *ecology* is usually credited to the German biologist Ernst Haeckel, who coined it in 1866 of two Greek words—*oikos*, meaning household or living relations, and *logos*, study of. He defined it as "the whole science of the relations of the organism to the environment including, in the broad sense, all the 'conditions of existence.' " Haeckel used the term in his efforts to interpret to the scientific world of Germany the significance of Charles Darwin's theory of natural selection and evolution and his concept of "the economy of nature," as presented in the *Origin of Species* (1859). To Darwin, rather than to Haeckel, belongs the principal credit for describing the complex functional interrelatedness of organisms and environment and the tendency of the evolutionary process to elaborate and diversify the biota to produce what ecologists today speak of as a system in dynamic equilibrium.[6]

The ecological implications of evolutionary thought were all but lost, however, in the furor over the very fact of evolution, including religious and social implications of the animal origins of man. Evolutionary research progressed along a number of discrete lines in various scientific disciplines, as did rudimentary "ecological" research (although without even the unifying value of a

6. For further discussion of Darwin's relationship to ecological thought, see Robert C. Stauffer, "Haeckel, Darwin, and Ecology," *Quarterly Review of Biology*, 32:2 (June 1957), 138–44; and "Ecology in the Long Manuscript Version of Darwin's *Origin of Species* and Linnaeus' *Oeconomy of Nature*," *Proceedings of the American Philosophical Society*, 104:2 (April 1960), 235–41.

common rubric, Haeckel's term *ecology* having failed to catch hold). Among the fields that began to develop environmental lines of investigation during the late nineteenth and early twentieth centuries were developmental and response physiology, hydrobiology, economic entomology, botany, and zoology; but many of the potentially most significant contributions to a modern functional ecology remained isolated, imbedded in the laboratories and literatures of the separate disciplines.

Ecology as a distinct scientific discipline is a product of the twentieth century. In the United States around the turn of the century it was plant ecology that gained attention and set the style for ensuing decades, with the work of Frederic E. Clements on "plant formations" and "climax" vegetation in the state of Nebraska and Henry C. Cowles' studies of vegetational succession on the sand dunes of Lake Michigan. Strongly influenced by the conceptual frameworks and investigative techniques developed by the Cowles and Clements schools of plant ecology, early animal ecologists like V. E. Shelford and C. C. Adams contented themselves largely with adding animals to the picture of succession. All of their approaches were primarily descriptive rather than functional.[7]

Thus ecology, as Aldo Leopold would first have encountered it, was a theoretical construct in botany, useful for describing vegetational patterns and related aspects of animal distribution. It was basically a subject matter, not a point of view, intriguing and no doubt exciting to those scientists who knew about it but still far from being a tool for integrating knowledge in a wide range of disciplines. Cultural geographers of the nineteenth century like George Perkins Marsh or natural philosophers like Henry David Thoreau or early conservationists such as John Muir, although in their own way intent on probing the interrela-

7. For the history of ecological science see W. C. Allee, A. E. Emerson, O. Park, T. Park, and K. P. Schmidt, *Principles of Animal Ecology* (Philadelphia, 1949), pp. 1–72; Richard Brewer, *A Brief History of Ecology: Part I—Pre-Nineteenth Century to 1919*, Occasional Papers of the C. C. Adams Center for Ecological Studies, No. 1 (Kalamazoo, Mich., 1960), 18 pp.; and Charles S. Elton, *The Pattern of Animal Communities* (London, 1966), pp. 29–44.

tionships of organism and environment, had scarcely heard of the term. The transformation of ecology from a descriptive schema in botany to a functional approach to the total environment—a concern with processes and relationships, with causes and effects—would occur in subsequent years, and Aldo Leopold would be involved in working out the implications of that transformation. At the start, however, it is probably safe to say that he was attracted more by his love of the outdoors and the excitement of the new conservation movement than by the intricacies of ecology.[8]

Aldo Leopold as Forester–Conservationist

Aldo Leopold was born in Burlington, Iowa, on January 11, 1887, the son of a prominent manufacturer of fine-quality walnut desks and grandson of a German-educated landscape architect who had designed a number of public buildings and parks in Burlington.[9] He grew up in a mansion atop a high, limestone bluff overlooking the

8. For a review of ecological science and scientists by a plant ecologist very critical of the dominant theoretical-descriptive approach and in favor of a holistic or functional approach see Frank E. Egler, "A Commentary on American Plant Ecology, Based on the Textbooks of 1947–1949," *Ecology*, 32 (Oct. 1951), 673–95. Among the few exceptions to his nearly universal derogation, Egler cited the promising new directions being explored by Norman Fassett, John T. Curtis, and Grant Cottam, all of the Botany Department at the University of Wisconsin. Fassett and Curtis were close personal friends of Leopold, and Cottam had been one of his students. Leopold, in an unpublished essay, "Conservation: In Whole or in Part?" holograph, 2 Nov. 1944, 8 pp., LP 6B18, discussed what he termed the *unity concept*, which seems roughly equivalent to Egler's *holism*.

9. Details and generalizations in this biographical sketch are based on the Leopold Papers in the U.W. Archives; Forest Service records in the National Archives, the Federal Records Centers in Denver and St. Louis, and various Forest Service offices; conversations with Mrs. Aldo Leopold, other members of the Leopold family, and colleagues, friends, and students of Aldo Leopold. Background details are drawn from some of the standard histories and policy studies in the field of natural resources, from the annual reports of various federal and state agencies, and from the Natural Resources History Project (tape recorded interviews) of the Wisconsin State Historical Society.

Mississippi River, where the thin, stony soil meant the family had to work unceasingly to encourage the array of wildflowers, trees, and shrubs they so enjoyed. Down the bluff and across the railroad tracks was the big river itself, migratory pathway for a quarter of the ducks and geese of the continent, its bottomlands a year-round wildlife wonderland for a growing boy. In those days there were no restrictions on hunting methods or seasons or bag limits, save those evolved in a personal code of the indivdual sportsman, and in later years Aldo Leopold recalled how his father had voluntarily quit shooting waterfowl in the spring, even though he still believed it was all right for his sons to shoot. During his school days in Burlington, at Lawrenceville Prep in New Jersey, and in Sheffield Scientific School at Yale, Leopold maintained a lively interest in field ornithology and natural history and began a lifelong practice of recording his observations daily in a journal.

In 1906 he began studies at Yale for a career in forestry, newest and most appealing of the outdoor professions. The possibilities for careers in forestry had been enhanced the year before when approximately 100 million acres of federal forest reserves were transferred from the jurisdiction of the Department of Interior to the Department of Agriculture for administration by the newly designated U.S. Forest Service, headed by Gifford Pinchot. The reserves had been set aside beginning in 1891 because of fear of impending timber shortage and the effects of forest destruction on water supply, but they had not been adequately administered. Pinchot's accomplishment was to win support for retention of the forests in public ownership "for the permanent good of the whole people," and to forge an organization of scientifically trained professionals capable of managing them for sustained production of timber, protection of watersheds, and grazing. The concept of sustained yield and wise use of resources, of management according to high standards of professionalism, efficiency, and public purpose, was the essence of the conservation idea espoused by President Theodore Roosevelt and elevated to the status of a national cause during the years when Aldo Leopold was studying forestry at Yale. The Yale Forest School, the first graduate school of forestry in the United States, had been established in 1900 with an endowment by the Pinchot family to provide a supply of "American foresters trained by Americans in American ways for the work ahead in American forests."

When Leopold graduated with a masters degree in June 1909, he joined the Forest Service and was sent to Arizona and New Mexico territories, where national forest administration was being organized in a new Southwestern District (District 3). That first summer, an utter greenhorn from the East, he headed a six-man reconnaissance party assigned to map and cruise timber in the wilderness fastness of the Blue Range, in the Apache National Forest of east-central Arizona. The seasoned locals and one Harvard tenderfoot on his crew were not as enamored as Leopold of roughing it on beans and biscuits; they did not appreciate his leaving them with the work while he went off exploring or chasing after Indians who were "making jerky"; and they considered his management of the technical reconnaissance as entirely incompetent.

But experience and promotions came fast in those days, and by 1912 Leopold was supervisor of the Carson National Forest in northern New Mexico, a million acres supporting 200,000 sheep, 7,000 head of cattle, 600 homesteads, and a billion feet of timber. He married Estella Bergere of a prominent old Spanish land-grant family, built a home near the forest at Tres Piedras, and reveled in the responsibilities of being a forest supervisor, to his mind far and away the most satisfying post in the Service. In April 1913, while settling a range dispute in a remote area of the forest, he was caught in a flood and then a blizzard and had to sleep in a wet bedroll. Within days an inflammation set in in his knees that was so severe he could not ride. A country physician, wrongly diagnosing it as rheumatism, prescribed the worst possible treatment, and by the time Leopold got to a doctor in Santa Fe he was bloated and near death, a victim of acute nephritis.

Eighteen long months of recuperation followed before he was strong enough to undertake even light office work. He had plenty of time in the interim to read, to fish, and to think. We are not certain what he read, but we can imagine he turned to the eleven-volume Riverside edition of Thoreau's works he had received as a wedding gift and to southwestern history and the narratives of the explorers and naturalists—Lewis and Clark, James O. Pattie, George F. Ruxton, Francis Parkman, John Burroughs, Ernest Thompson Seton—which fascinated him all his life. One of his colleagues on the Carson, Raymond Marsh, has suggested that these months of enforced inactivity and contemplation marked a decisive change in Leopold's outlook. By the time he returned to work as acting head of the office of grazing at the headquarters of District 3 in Albuquerque, he was beginning to realize there was no hope of resuming the strenuous, glorious field existence of a forest supervisor. A recurrence of the disease, which could be brought on by overexertion, was considered in all cases fatal. It was at this juncture that Leopold became involved in wildlife conservation work.

Americans had so depleted their stock of native wildlife by indiscriminate hunting, whether for market or sport, and in some instances by wholesale destruction of habitat, that by the late nineteenth century many species were in imminent danger of extinction and the future of sport hunting appeared bleak indeed. Certain segments of the public, notably sportsmen from the eastern states, had begun to organize to promote stricter game laws and enforcement, abolition of market hunting, and creation of game preserves for threatened species. In 1887 Theodore Roosevelt organized the Boone and Crockett Club from a select group of politically well-placed big-game hunters; in 1902 the National Association of Audubon Societies was established; William T. Hornaday founded his Permanent Wildlife Protection Fund during 1910–1912; and in 1911, on a somewhat different tack, the American Game Protective and Propagation Association was founded, with partial funding from sporting arms and ammunition manufacturers, to begin developing scientifically grounded wildlife conservation programs. During this period practically all the states established some sort of fish and game administration, although these agencies tended at first to be inadequately staffed by political appointees.

In the southwestern mountains, which did not attract

substantial Anglo-American settlement until after about 1885, the scarcity of wildlife was just beginning to be noticed when Aldo Leopold arrived on the scene. Although most of the remaining game animals, especially deer and turkeys, were on national forest lands, the Forest Service had no legislative mandate to administer its lands for wildlife or recreation or indeed for anything but timber production and watershed values.[10] In the case of wildlife, an added problem involved jurisdiction. Under the tradition of English common law dating back to the Magna Carta, wildlife was regarded as the property of the people as a whole. Historically, it fell to the jurisdiction of the several American colonies and subsequently to the states, rather than to the private landowner, as in the continental European system, or to the federal government (which even today owns a third of the land area of the nation). But the new states of Arizona and New Mexico, admitted to the Union in 1912, did not have enough game wardens effectively to patrol the vast, roadless acreages of the forests. Ever alert to strategic opportunities for building a constituency that would support federal retention and management of the forests, the Forest Service quickly concluded cooperative agreements with Arizona and New Mexico, under which forest officers would be deputized to help enforce state game laws. Rangers were on the ground anyway and could apprehend violators while performing their other duties. So went the theory, but in practice not a single arrest had been made up to the time Aldo Leopold became involved in 1915.

Leopold may have been responsible for overseeing the cooperative agreements as acting head of the office of grazing. In any event, by June 1915 he managed to get himself assigned almost full time to organizing game and fish work in the Southwestern District. He immediately prepared a mimeographed "Game and Fish Handbook," explicitly defining the duties and powers of forest officers in coopera-

10. The congressional mandate for preservation of timber and watershed values dates from the Sundry Civil Appropriations Act of June 4, 1897. Equal status for recreation, range, wildlife, and fish purposes had to await the Multiple Use-Sustained Yield Act of 1960.

tive game work, which attracted favorable attention in Washington and in other offices of the Forest Service around the country. In October he was host to Dr. William T. Hornaday, director of the New York Zoological Park, president of the Permanent Wildlife Protection Fund, prolific writer, and longtime crusader in the cause of wildlife conservation, who spent several days in Albuquerque on a western tour to drum up support for the "Hornaday Plan" for national forest wildlife refuges. Hornaday had evolved from an avid hunter to a strict protectionist, bitterly opposed to most hunting whether for meat or for sport. He could be notoriously caustic toward fellow conservationists who were less extreme than himself, but there was no question about his ability to muster public sentiment.

Whether inspired by Hornaday or by his own consciousness of needs and opportunities, Leopold devoted the next few months to stumping the district. He met with local forest officers and citizens to organize local game protective associations and promoted strict enforcement of game laws, eradication of predatory animals, creation of game refuges, and restocking of depleted lands and waters. His unabashed use of the term *game* rather than *wildlife* was itself evidence of his commitment to perpetuate sport hunting, W. T. Hornaday notwithstanding. Extraordinarily persuasive in personal contact, he proved himself a master at appealing to diverse interest groups. He spoke not only to sportsmen and foresters but to businessmen, for whom he painted a glowing picture of the region's potential as the most valuable vacation ground in the world, to ranchers, whom he attracted to the refuge idea by suggesting the prospect of securing the cooperation of sportsmen in the war against predators, and even to Indians in pueblos bordering the forests. We have his own description of his feats, prepared for his Yale class record, in which his ardor for the cause overcame his modesty:

> We have about twenty million acres of Forest in this District, part of which is unfit for livestock, and on these waste lands I ultimately plan to raise enough game and fish to provide recreation for twenty thousand people and

bring $25,000,000 a year into the country. This is an ambitious project but I know it can be done and I have got the public to where they are about ready to believe me. I am organizing game protective associations over both states, securing the reintroduction of locally extinct species, stocking hundreds of waters with trout, fighting suits for violation of the game laws, giving illustrated lectures to the public, hammering on game protection through the newspapers, raising a fight on predatory animals, and have written a book outlining plans, ways, and means. While making good progress I think the job will last me the rest of my life.[11]

In the midst of this activity, late in 1915, Leopold was asked to accept a detail as editorial assistant in the Washington office of the Forest Service. Not wanting to leave the game protection movement before it was ready to stand on its own, he declined the assignment, only to be officially ordered by the Chief Forester to accept it. He responded with a long letter explaining that because of the uncertainties of his health and his inability ever again to do strenuous field work, he had to watch his opportunities carefully if he did not want to end up in a "dead" job like information work. "To speak plainly," he added in a handwritten note, "I do not know whether I have twenty days or twenty years ahead of me. Whatever time I may have, I wish to accomplish something definite." For him, that something was obviously game protection. As long as he could work on game protection through the Forest Service he preferred that position, but now it seemed he was forced to choose between the Service and doing his chosen work. "To abandon a chance at a live field in favor of a sure job at nothing-at-all," he concluded, "would be playing quitter."[12]

11. "For the 1908S. Class Record, Yale University," holograph, c. 1916, 5 pp., Aldo Leopold folder, Sheffield Scientific School, Yale University.

12. AL to A. C. Ringland, 14 Feb. 1916, LP 11M1 (original at Federal Records Center, St. Louis). Leopold was on a protein-free diet during these years. He got regular rest and was extra careful not to overexert himself for fear of a fatal relapse. By the 1920s, however, he was fully recovered and no longer worried about his health, according to Mrs. Leopold.

The order was changed, and Leopold stayed with the Service in Albuquerque. His zealous efforts in an unconventional field and the amount of public attention they attracted might well have been disturbing to traditionally utilitarian foresters. Indeed the Service remained reluctant to commit men and money to the multifaceted program of game restoration that Leopold envisioned. Yet ranking officers in District 3 seem to have gloried in his organizing feats, perhaps because they believed his work was strengthening the position of the Forest Service in its nascent struggle with the National Park Service.

Establishment of the Park Service in the Department of Interior in 1916 initiated decades of perennial behind-the-scenes jockeying between Agriculture and Interior for control of recreational lands. Anything the Forest Service could do to demonstrate potential for forms of recreation that were compatible with other more economic uses of the forests might help it to maintain control over prime lands coveted by Interior for new national parks. Hunting was one such form of recreation not provided for in the parks. Another was leased sites for summer homes and commercial recreational establishments, authorized in the Agricultural Appropriations Act of March 4, 1915, an act that marked the first significant congressional recognition of recreation as a legitimate use of the forests.

It was Leopold who was charged with planning recreational uses in the Southwestern District. After his initial splurge of activity in game protection, he had to devote his time increasingly during 1916–1917 to consulting with local forest officers on recreational working plans, laying out homesites and public campgrounds, developing private and commercial leasing policy, devising adequate sanitation facilities and regulations, and preparing promotional literature. His reluctance to see certain areas subdivided for recreational "improvements" would lead him, in a few years, to promote yet another substantial innovation in the recreation policy of the Forest Service, establishment of a system of roadless wilderness areas.

But game conservation remained a goal for which he continued to press, both within the Service and as an all-absorbing hobby in his spare time—that is, when he was not down on the Rio Grande with his four-year-old son Starker, shooting doves and ducks. He published articles on game conservation, forest policy, and ornithological observations; he started a number of personal research projects; and he became involved, through wide-ranging national contacts, in battles for federal legislation dealing with refuges and migratory waterfowl. As secretary of the New Mexico Game Protective Association he edited its official bulletin, *The Pine Cone*, a quarterly newspaper he had founded in December 1915 as an oracle of the new movement, and through it spearheaded the drive for a nonpolitical commission form of state conservation administration. His role in "the awakening of New Mexico" won him the gold medal of W. T. Hornaday's Permanent Wildlife Protection Fund in 1917 and a special commendation from Theodore Roosevelt, not to mention attractive job offers back East, which he declined.

When the stringencies of World War I forced discontinuance of virtually all game work in the Southwestern District, Leopold actually left the Forest Service in January 1918 to take a position as secretary of the Albuquerque Chamber of Commerce, where he hoped he could more effectively promote the cause of game conservation. Through the chamber he also promoted victory gardens, drainage of the Rio Grande Valley for agriculture, public parks, and a civic center for Albuquerque, with indigenous Spanish architecture.

Commerce was not all conservation, however, and the Forest Service was a compelling institution. In 1919 he returned to be assistant district forester in charge of operations, a position that entailed responsibility for business organization, personnel, finance, roads and trails, and fire control on the twenty million acres of national forests in the Southwest. It was as an administrator, not as a land manager, scientific researcher, or conservationist, that he made his mark in the next five years. Yet along with his very real accomplishments in developing more efficient personnel practices, fire control procedures, and methods for inspecting forests went a deep and active commitment to other less traditional concerns of the Forest Service, such as watersheds, wildlife, and wilderness. He helped stimulate research on erosion control and prepared a watershed handbook for the district. As he crisscrossed the forests on inspection trips he made notes on wildlife species and habitat conditions for a book on southwestern game that he intended to write with two sportsmen—colleagues. Not least important was the groundwork he laid for administrative designation in 1924 of more than a half-million acres in the Gila National Forest as wilderness, setting the pattern for the system of roadless wilderness areas that was given force of law in the National Wilderness Preservation Act of 1964.[13]

Aldo Leopold's writings during his years in the Southwest reveal both his enthusiasm for the conservation idea and his evolving awareness of environmental interrelationships. He applied the concepts of wise use and sustained yield to game conservation as well as forestry and in writings on both game and forestry sought to develop standards of skill and efficiency in the management of resources. His concern for conservation began to merge with

13. Leopold has been regarded as the "father" of the national wilderness system, although his claim to the title has been disputed by Donald N. Baldwin in "Wilderness: Concept and Challenge," *Colorado Magazine*, 44:3 (Summer 1967), 224–40, and in *The Quiet Revolution: Grass Roots of Today's Wilderness Preservation Movement* (Boulder, Colo., 1972). Baldwin would give the honor to Arthur Carhart, a landscape architect with the Rocky Mountain District, who discussed the wilderness area concept with Leopold in 1919 and won administrative approval for recreation plans recommending wilderness-type management for Trappers Lake in Colorado and the Superior National Forest canoe country in Minnesota in advance of Leopold's wilderness proposal for the Gila. The Forest Service, however, in celebrating the fiftieth anniversary of wilderness at the Gila National Forest in 1974 declared that the Gila had been the first area officially designated as wilderness. In any event, Leopold, as one of the elite corps of Yale-educated foresters who dominated the Service for decades, seems to have had a more influential role than Carhart in formulating wilderness concepts and policies on a national scale. The Wilderness Act of 1964 legally enforced the reservation of more than 9 million acres of administratively designated wilderness, protection for which Leopold had argued at least since 1925, and provided a review procedure for another 5 million acres of primitive areas in national forests and for some millions of acres of national park and national wildlife refuge lands.

his interest in ecological science as he searched for criteria of environmental quality in southwestern forests, watersheds, and rangelands and probed for a more comprehensive philosophy on the relations of man and environment.

Although Leopold was unquestionably familiar with the concepts of plant ecology emanating from the universities of Chicago and Nebraska, particularly as they described the distribution and succession of vegetation types, his thinking does not seem to have been dominated by them. Every bit as important as origins for his ideas about the southwestern environment were his habit of keen observation and his historical curiosity, coupled with his voracious reading of the great naturalists and the journals of the early explorers. In fact, Leopold may be said to have been thinking ecologically, in the functional or holistic sense, before ecological science had evolved a conceptual framework capable of supporting such thought.

While leading plant ecologists still described normal successional stages as a response to average environmental factors, Leopold through careful observation and inferential reasoning arrived at an essentially functional interpretation of vegetation change and soil erosion on southwestern watersheds—an interpretation that integrated soils, vegetation, topography and climate, geologic and human history, lightning fires and livestock grazing into a single system of interactions. Notably deficient in his early interpretation, on retrospect, was the wildlife component. Although he was among the earliest to appreciate the extent to which wildlife populations were limited by environmental factors, factors that could be manipulated to achieve greater production or control of the game resource, he did not yet view wildlife in its functional interrelations with the total land community. Yet this detracts hardly at all from the scope of his achievement. Leopold was left with a profound respect for the fragile equilibrium of the arid Southwest, in which man's activities in one part of the system were capable of inducing massive, sometimes progressive, usually unanticipated, and too often unrecognized changes in other parts of the system. It was an environment set, as he termed it, on "hair-trigger." As a conservationist, he was concerned with the implications of his interpretation for human action. Action involved changes not only in patterns of land use, but also in institutional arrangements affecting land use and, even more fundamentally, in the perceptions, attitudes, and values of a people. Leopold saw all this at least as early as 1923 and expressed it in a manuscript, "Some Fundamentals of Conservation in the Southwest."

Casting about for philosophical underpinnings for his interpretation of the hair-trigger equilibrium in the Southwest, he discovered the organicism of the Russian philosopher P. D. Ouspensky, who regarded the whole earth and the smallest particle thereof as a living being, possessed of soul or consciousness. "Possibly, in our intuitive perceptions, which may be truer than our science and less impeded by words than our philosophies," Leopold wrote, "we realize the indivisibility of the earth—its soil, mountains, rivers, forests, climate, plants, and animals, and respect it collectively not only as a useful servant but as a living being." Leopold never published this first attempt at formulating a philosophy of conservation. In later years, as ecological science became more functional and holistic, he would begin to couch his land ethic in ecological concepts rather than in the terminology of the philosophers.[14]

The Wisconsin Years

Just as his administrative, scientific, and philosophical concerns were converging in his southwestern environment, in 1924 Leopold was asked to accept a transfer to the U.S. Forest Products Laboratory in Madison, Wisconsin. The Forest Products Laboratory was the principal research institution of the Forest Service at that time, but the nearly 150 chemists, engineers, physicists, mechanics,

14. "Some Fundamentals of Conservation in the Southwest," typescript, c. 1923, 19 pp., LP 6B16. How or in what form Leopold stumbled on Ouspensky remains a mystery. Ouspensky is one of the few philosophers Leopold ever quoted in his writings. In addition to this reference, he mentioned Ouspensky (or simply "a Russian philosopher") about three times in connection with the terminology of *phenomenon–numenon*. See for example p. 3, fn. 3.

and foresters on its staff were concerned almost exclusively with research on forest products rather than with the growing of trees. No doubt because of his proven skill as an administrator and his sympathetic understanding of a broad range of research needs, Leopold was offered the position of associate director, with the understanding that there was "more than a possibility" that he would become director within a year. Moreover, he would have the encouragement and support of the Washington office to move the activities of the laboratory into closer correlation with the field units of the Service and the whole forest conservation movement. Having rejected at least five previous offers for promotion that would have entailed leaving the Southwest, he accepted the position at the laboratory and reluctantly moved his family. But the incumbent director did not resign, and Leopold spent four frustrating years in the number two slot fighting a mountainous administrative load. He tried to spur interest in utilization of waste wood and inferior species, in genetic and site research for improved quality of trees, and in other aspects of laboratory–field cooperation—all to devastatingly little avail. One can imagine that he felt constrained in an institution whose primary concern was with utilization of the tree after it was cut, when everything about him made him interested in the forest as a living community.

He made what shifts he could to function effectively in the laboratory setting, but his principal release during those years must have come from his hobbies. His primary interest during his spare time was working on his manuscript, "Southwestern Game Fields." The book was to include life histories of southwestern wildlife species and an illustration of the principles of game management as applied to a single species, deer, in a particular area, the Gila wilderness. He prepared drafts of various chapters for two different versions of the book and circulated them among his colleagues in New Mexico for comments, but an unanticipated irruption of deer on the wolfless Gila coupled with his own distance from the scene forced him, in the end, to abandon the manuscript. Despite the time he spent writing, Leopold did get out a good deal on weekends into the countryside around Madison. He also managed to pick up a new hobby that infected the whole family—archery, for which he made his own bows and arrows and even his glues. The new avocation justified several return visits to the Southwest to try his luck on the superabundant deer of the Gila.

Leopold also continued his involvement in conservation politics. No sooner did he arrive in Wisconsin in 1924 than he was swept into the local chapter of the Izaak Walton League, recently organized by an able and spirited group of citizens to promote a stronger state forestry program and a more effective, less political state conservation administration. Having had experience drafting and promoting a proposal for a state game commission in New Mexico, enacted in 1921, he was a key figure in the effort that culminated in Wisconsin's Conservation Act of 1927. The act provided for a conservation department headed by a director responsible to six unpaid commissioners, appointed by the governor for staggered six-year terms. Unlike the New Mexico commission, which was concerned only with fish, game, and enforcement of laws that pertained to them, the Wisconsin commission was responsible for forests as well.

It should be noted that eastern states like Wisconsin had no system of national forests to fall back on, most of the public domain in the East having been taken up by private interests before the federal forest reserves were established. Wisconsin's magnificent white pine forests had been virtually mowed down in the westward march of the timber barons during the period 1870–1910. The Weeks Law of 1911 provided for federal repurchase of cutover forest lands in the eastern states, and acquisition was begun under this law in the early 1930s for two national forests in Wisconsin, the Chequamegon and the Nicolet. But the bulk of public forest acreage in Wisconsin, nearly all of it cutover, tax-reverted land, was acquired by the state or the counties under a cluster of enabling laws passed in the 1920s and administered by the new conservation commission.

With his professional background in forestry and his personal interest in game management, Leopold was the natural candidate for director of the conservation department under the new commission, so he and his Waltonian cohorts thought, and he was prepared to leave his position with the Forest Products Laboratory as soon as the appointment could be secured. But the governor and his appointed commissioners did not oblige, and Leopold was to experience more than a decade of bitter frustration in his repeated attempts to cooperate with the conservation administration he helped create.

By 1928, however, Leopold was determined to leave the laboratory for a position more in line with his consuming interests in wildlife and conservation. Declining more secure opportunities with the Forest Service and various universities, he chose to strike off on his own into a new profession, game management. Under funding from the Sporting Arms and Ammunition Manufacturers' Institute, he began conducting game surveys of the north-central states. A game survey, as Leopold envisioned it, was the first step in game management; it involved appraising the environmental factors affecting productivity of game in a particular region and recommending policy measures necessary to restore game. It was also "an attempt to change the orientation of thought and action on wildlife conservation"—to show, in terms of local conditions and practices, the difference between the old idea of restricting the kill and the new idea of building up the supply through management of habitat. From July 1928 to January 1930 Leopold traveled through Michigan, Iowa, Minnesota, Ohio, Mississippi, Illinois, Indiana, Wisconsin, and Missouri, spending two weeks to two months in each state, visiting more than three hundred localities and consulting with more than six hundred state and local officials, scientists, and sportsmen. He prepared typewritten reports, charts, and maps for each state and summarized his findings in his *Report on a Game Surey of the North Central States,* published early in 1931. With funding from the institute he also set up a series of game research fellowships at five universities and delivered a course of lectures on game management at the University of Wisconsin.

The game survey and related work, coupled with his earlier activities in the Southwest, established Leopold as one of the country's foremost authorities on native game.

Shortly after beginning the survey, he became chairman and chief draftsman of a committee charged with formulating a policy concerning game in America. Adopted by the Seventeenth American Game Conference in 1930, the game policy signaled a new approach to wildlife conservation in the United States. Up to that point the emphasis had been either on crusading, in the manner of William T. Hornaday, for hunting laws, refuges, and other devices to preserve remnants of diminishing species, or on artificial propagation by game breeders, gun club operators, and state game departments. The new policy, like Leopold's game survey, stressed the idea of production in the wild. It advocated encouragement of management of habitat *by the landholder*, whether public or private, forester, farmer, or weekend recreationist; and it encouraged experimentation in various methods of cooperation among landowners, sportsmen, and the nonshooting public. Game production was not a matter of witch-doctoring or abstract theorizing. With all its emphasis on experimentation, the policy also stressed the need for solid scientific foundations and the training of men for administration, management, and research. In short, it was necessary to "make game a profession."

Aldo Leopold is acknowledged the "father" of the profession of wildlife management in America. One man can hardly establish a profession, but Leopold's stamp has been on the profession so conspicuously from its beginnings around 1930 to the present that the title is perhaps justified. Professionalism in his estimation was a matter not so much of academic degrees as of point of view, technical understanding, standards, and skill. Yet to secure these professional attributes on a broad scale required a variety of institutional arrangements for the conduct and application of research, for the development of specialized vocational, technical, and scientific training, and for the establishment and maintenance of high standards of technical performance and ethical conduct. Others were teaching and working in the wildlife field and even in wildlife management before Leopold, but no one saw more clearly than he the need for sound institutional foundations—in the universities, in government agencies, and in private organizations—nor worked more effectively to create them. In this he had undoubtedly been inspired by the example of forestry, which had emerged as an area of employment in the earliest years of the century and within decades had developed the institutional structure to support a complete transformation to a professional basis at the field level, in the U.S. Forest Service at least. At first, Leopold had called upon his colleagues in the forestry profession to develop a science of game management; but the turning point in his career came in 1928 when he left the Forest Products Laboratory to begin laying the foundations for a distinct new profession.

It takes confidence in one's own abilities and faith in the future to leave the security of an established institution like the U.S. Forest Service and strike out, midway in one's career, into a profession not yet born. But for Leopold with his compelling desire to *build* something, with his life as well as on the land, it was a characteristic move. As it happened, the stock-market crash of October 1929 and the ensuing depression knocked out his funding from the arms manufacturers by 1931, leaving him with a wife and five children to support and a letterhead proclaiming his availability as a consulting forester. He did manage to pick up a few months' work conducting a second game survey of Iowa and a survey of potential game management areas in Wisconsin. But nothing better illustrates his optimism and commitment than the calm discipline with which he applied himself, unemployed during what for most Americans were bewildering, hopeless years, to writing a textbook for the new field. Based on the most recent developments in wildlife research and permeated with Leopold's rare esthetic and philosophic sense, *Game Management* (Charles Scribner's Sons, 1933) is still regarded as a basic statement of the science, art, and profession of wildlife management. It has been continuously in print since 1933 and makes fascinating reading for the layman as well as the professional.

In writing *Game Management* Leopold not only utilized his own manuscript on southwestern game, the findings of the game survey, and his lectures at the University of Wisconsin, but also drew on the work of his predecessors and colleagues in the field of wildlife. Unquestionably the most significant early research in game management was by Herbert L. Stoddard of the U.S. Biological Survey, who in 1924 had begun an investigation of quail populations in Georgia in cooperation with the owners of huge, private quail preserves. Stoddard made important findings on the role of fire in maintaining favorable quail habitat and productive timber stands and on the function of predation in adjusting population levels and promoting vigorous stock; and he produced a classic life history and management study, *The Bobwhite Quail* (1931). Leopold drew also, perhaps more than he realized or acknowledged, on the practical experience of another extraordinarily able wildlife manager, Wallace Byron Grange. Grange had been the first superintendent of game for the Wisconsin Conservation Department and would later author another classic in the field of wildlife management, *The Way to Game Abundance* (1949). Stoddard, Grange, and Leopold worked together after 1928 supervising the Sporting Arms wildlife research fellowships, which supported significant research on upland game birds. Most notable was the work of Paul Errington at the University of Wisconsin on the relationship of predation to environmental carrying capacity and population density in the northern bobwhite.

Although Leopold relied on the techniques and findings of Stoddard, Grange, Errington, and other field researchers and on his own previous work for much of the substance of his book, he cast much of the material in terms of ecological concepts being formulated by still other scientists, including the eminent British ecologist, Charles Elton. Elton, whose first major work, *Animal Ecology* (1927), signaled the beginning of a gradual shift from a primarily descriptive to a functional approach in ecology, was one of the first to employ the concept of ecological niche, in the sense of the functional status of an organism in its community, and he elaborated the concept of food chains as the basic organizing principle of the community. Leopold had met Elton in 1931 at the Matamek Conference on Biological Cycles, and the two had struck up an immediate and enduring friendship. That Leopold by 1933 should have integrated the new functional concepts of ecology so well with field observations and research in game management as to produce a book that is still regarded as a classic in the field is unquestionably a remarkable accomplishment. Yet it should be noted that the word *ecology* scarcely appears in the text, and there is but one reference in the index to *ecologic niche*. The compelling idea for Leopold in 1933 was not the idea of ecology so much as the idea of management.

Management, the art of producing sustained yields of wild game, had been the key to his efforts almost since he first became involved in game conservation. Corollary to management was the idea of control, which he defined in *Game Management* as "the coordination of science and use." "The central thesis of game management," he said in his preface, "is this: game can be restored by the *creative use* of the same tools which have heretofore destroyed it—axe, plow, cow, fire, and gun." Management was the purposeful and continuing alignment, or control, of these forces. In his emphasis on management, Leopold simply extended to wildlife, through the medium of rudimentary ecological science, a faith in the possibility of intelligent control that goes back at least to W G McGee, Gifford Pinchot, and the origins of the conservation movement in America.

Leopold's faith in the idea of management, conceived as control, extended to the environment of man as well as of game. "I will not belabor the pipedream," he told the Southwestern Association for the Advancement of Science in a major address, "The Conservation Ethic," in May 1933. "It is no prediction, but merely an assertion that the idea of controlled environment contains colors and brushes wherewith society may some day paint a new and possibly a better picture of itself." The economic cards, especially in the depths of a depression, seemed to be stacked against the most important reforms in land use. But permanent though economic laws may be, Leopold pointed out, "their impact reflects what people want, which in turn reflects what they know and what they are." His was a plea for ecological understanding, for the extension of ethics from the realm of human social relations to the whole land community of which man was an interdependent member. But

again, as in *Game Management,* the emphasis was not so
much on the concepts of ecology as on the use of tools—
tools economic, legal and political, as well as scientific and
technical—to create a more enduring civilization.[15]

An opportunity to try out some of his ideas about management was not long in coming. In August 1933 a chair
of game management was created for Leopold in the Department of Agricultural Economics at the University of
Wisconsin. Supported by an unprecedented five-year grant
from the Wisconsin Alumni Research Foundation, the
chair could be justified in the midst of a depression by
its potential contributions in the realm of land utilization
—development of a productive game crop—on Wisconsin's
cutover, tax-reverted, burned-out, and eroded lands; hence
the rationale for placing it with agricultural economics.
Leopold had been angling for a position at the university
for years, and he would remain there for the rest of his
life. (A one-man Department of Wildlife Management was
established by the university in 1939.) He set up a small
graduate-training operation and established a number of
farm demonstration areas near Madison where he and
his students could experiment with cooperative farmer–
sportsman arrangements, get practical land-management
experience, and conduct field research on wildlife. He
served also as research director of the newly established
University of Wisconsin Arboretum, working with professors and students from various disciplines to plan and
conduct the restoration of native ecological communities.

Though he may have thought he would be left alone to
concentrate on working out his ideas in his own little
corner of the country, events in Washington drew him
abruptly into problems and programs on the national
scene and dramatized, almost immediately, both the necessity and the difficulty of what he was trying to do. Leopold's chair was established just as President Franklin D.
Roosevelt's New Deal was shifting into high gear and
millions of federal dollars suddenly became available for
work-relief projects and purchase of sub-marginal lands,
under AAA, CCC, FERA, WPA, ECW, SES, and other

15. "The Conservation Ethic," *Journal of Forestry,* 31:6
(Oct. 1933), 634–43.

alphabetical apparitions. Inauguration of all these programs, each with at least a potential wildlife component, generated an extraordinary demand for trained supervisory personnel, a demand which would obviously be met because there was money, but not necessarily met well. Leopold placed a few of his students in technical field positions with federal agencies and himself served as advisor to a number of conservation projects in Wisconsin, stressing in each case the need for cooperative integration of land uses, including farming, forestry, wildlife, and recreation, and the need to tailor programs to local conditions and to involve individual landholders. Thousands of men were clamoring for jobs, however, and money was waiting to be spent. The inevitable result was new roads, trails, ditches, dams, wherever and as soon as they could be built. The more he saw the more disillusioned he became about the prospects for ever achieving integrated conservation from the disjointed functioning of single-track relief agencies.

In early 1934 he served with the cartoonist J. N. (Ding) Darling of Iowa and Thomas Beck of Collier's Publishing Company on the President's Committee on Wildlife Restoration. They were charged with drafting a proposal for dovetailing Roosevelt's $25 million program for federal purchase of submarginal farmland with a program for restoration of wildlife habitat. Leopold stood alone on the committee in arguing for more federal encouragement of research and administrative coordination by the states, which he thought were in a better position than the federal government to deal with local conditions and foster the practice of game management by private landowners. When he was asked several months later to take over as Chief of the U.S. Biological Survey, the federal agency responsible for implementing the program for wildlife restoration, he declined. He was interested more in research and demonstration than in land acquisition, and he thought it might prove just as important in the long run for him to bring research to actual fruition in Wisconsin as to try his hand at starting it nationwide, especially when there were as yet no federal funds in sight for research.

Through his continuing contacts with federal officials and through scientific and professional societies, he continued to press for establishment of research programs, especially at the state level. His efforts were rewarded in 1935 with creation of the Cooperative Wildlife Research Unit Program, which provided for research units in nine land-grant colleges across the nation. But to his bitter disappointment his own university, pioneering institution in the field of game management, failed to get one of the units because the Wisconsin Conservation Commission refused to cooperate.

From the start, Leopold had been limping along with a dearth of research funds, less than $4,000 a year, to support all his students. Yet he resisted the temptation to push students quickly through the mill and out into the burgeoning federal agencies. Rather, he insisted that they attain a solid foundation in a wide array of related disciplines, acquire actual field experience in technical game management, and carry out a well-conceived, publishable research project, at the masters as well as the doctoral level. He took only as many students as he had time to work with individually, selecting them for their promise in the new field rather than for their past record. Among his students, who are now almost without exception leaders in the fields of wildlife or natural resource management, Leopold was and still is known simply as "the professor," a designation he always cherished.

Toward an Ecological Philosophy

Many of Leopold's early students have remarked at his youthful, inquiring mind, his openness to new ideas, and his willingness to move in new directions. They had ample demonstration of these qualities in the mid-1930s, for it soon became apparent that control of game populations would be more difficult than they had thought. They had started at both ends at once, doing research on the life histories of various species and putting in food patches and cover plantings on the demonstration areas in an effort to build up populations. Key species like quail and grouse, however, on which they had concentrated most of their efforts, failed to increase as expected and instead oscillated in response to some unknown cause. Elsewhere in the country deer herds mushroomed all out of control, and game officials were finding it nearly impossible, in the absence of solid, research-based fact, to win public support for adequate reduction. Hence Leopold and his students found themselves moving increasingly in the direction of more basic ecological research on animal population mechanisms—"deep-digging" research, Leopold called it—and putting less trust in simple manipulations of habitat.

The move toward more basic ecological research in the mid-1930s was more than a quest for new facts or relationships. A close analysis of Leopold's writings in these years reveals a subtle though highly significant shift in his whole intellectual orientation, a shift somehow symbolized by three events in his life in the year 1935. In January he joined with Robert Marshall and others to found the Wilderness Society, a national organization to protect and extend the increasingly vulnerable system of wilderness areas they had been instrumental in creating. For Leopold the new society had philosophical as well as political significance. It was "one of the focal points of a new attitude—an intelligent humility toward man's place in nature." This new attitude involved a commitment to preserve threatened species, especially predatory animals such as wolves and grizzlies, which Leopold now realized were essential to the healthy functioning of ecosystems. The year 1935 marked a reorientation in his thinking from a historical and recreational to a predominantly ecological and ethical justification for wilderness.[16]

In April, Leopold acquired the worn-out, abandoned farm on the Wisconsin River that was to become the setting for most of the nature sketches in *A Sand County Almanac.* "The shack," as the Leopold family fondly dubbed the old chickenhouse they refashioned into essential lodgings, became weekend and vacation headquarters for the soul-satisfying experience of restoring the land to ecological integrity.

And in autumn of 1935 he spent three months in Ger-

16. "Why the Wilderness Society?" *The Living Wilderness,* no. 1 (Sept. 1935), 6.

many on a Carl Schurz traveling fellowship, studying German methods in forestry and wildlife management. It was his first and only trip abroad and an eye-opening experience. His confrontation with the ecological and esthetic costs of the highly artificialized German system of management, particularly with respect to deer and forests, challenged some of his most basic assumptions about the ultimate possibility of environmental control and led him to reevaluate the objectives of wildlife management. No single event can cause a transformation in the intellectual development of so integral a thinker as Leopold, but surely the impact of the German experience, his redefinition of the wilderness idea, and the convergence of observation, activity, and reflection at his sand-country shack signal in important ways the beginnings of his mature philosophy.

The mid-1930s were significant years also in the biological sciences, especially in the realms of ecological and evolutionary theory. Ernst Mayr in his monumental *Animal Species and Evolution* (1966) identifies the 1930s as the period when the various discrete lines of specialization in evolutionary biology were "almost suddenly fused" into a broad unified theory. And "the great AEPPS" (Allee, Emerson, Park, Park, and Schmidt, *Principles of Animal Ecology*, 1949) cite these years as a time of acute interest in theoretical ecology and ecological aspects of evolution. Evolution and ecology were coming to be recognized as two windows on the same process. Developments in the biological sciences undoubtedly helped Leopold conceptualize his new approach to wildlife and land management, but one gains the impression from his writings that his experiences in game management, his trip to Germany, and his activity at the shack were more instrumental in effecting the transformation in his thinking.

Leopold's earliest comprehensive statement of the new ecological viewpoint was his paper, "A Biotic View of Land," read in June 1939 before a joint meeting of the Society of American Foresters and the Ecological Society of America. Here he first presented the image of land as a biotic pyramid—"a fountain of energy flowing through a circuit of soils, plants and animals"—and drew ecological interrelationships into an evolutionary context. The whole trend of evolution, he suggested, was to elaborate and diversify the biota, to add layer upon layer to the pyramid, link after link to the food chains of which it was composed. He asserted, further, that the normal circulation of energy among the various levels of the pyramid—the stability or healthy functioning of the system—depended on the complex structure of the whole, much as the upward flow of sap in a tree depends on its complex cellular organization. Structure, he pointed out, was manifest in the characteristic numbers as well as the characteristic kinds and functions of species. The old approach of economic biology that conceived of the biota as a system of competitions and sought to give a competitive advantage to those species deemed "useful," whether corn or pines or deer, as against those deemed harmful or expendable, would have to give way to a new ecological approach which conceived of the biota as a single system, the land organism, "so complex, so conditioned by interwoven cooperations and competitions, that no man can say where utility begins or ends." Thus did Leopold express the transition from conservation as a preoccupation with supply and demand to conservation as a state of land health.[17]

The key idea in this essay was Leopold's assumption that there was a definite relationship between the complex structure and the smooth functioning of the whole—between the evolution of ecological diversity and the capacity of the land organism for self-renewal, which he termed *stability* or *land health*. As testimony to the crucial role that this apparent relationship between diversity and stability assumed in Leopold's mature thinking, we have a number of unfinished manuscripts and manuscript fragments in which he wrote of "circumstantial evidence," "the tacit evidence of evolution," and even "an act of faith." Indeed, such a relationship is not proven even today, although we can be sure that it is no simple relationship.[18]

17. "A Biotic View of Land," *Journal of Forestry*, 37:9 (Sept. 1939), 727–30.
18. See manuscripts in LP 6B16 and 6B18. Charles Elton has reviewed scientific evidence suggestive of a relationship between diversity and stability in *The Ecology of Invasions by Animals and Plants* (London, 1958).

The objective of conservation in a system thus understood was to preserve the capacity for healthy functioning of the system, rather than primarily to protect individual animals, à la Hornaday, or to produce a shootable surplus, as in early game management. Three decades of experience trying to "control" wildlife populations by manipulating selected environmental factors had had a profoundly sobering effect on Leopold. A proper function of management, it now became apparent to him, was to encourage the greatest possible diversity in an attempt to preserve the widest possible realm in which natural processes might seek their own equilibrium.[19]

Along with Leopold's greater consciousness of ecological enigmas and of the necessity for "deep-digging" research came an impatience with the prevailing emphasis of wildlife managers and government agencies on practicality and their insistence on "blood-and-feathers dividends." Although the name of the profession had changed in less than a decade from the rather too economic *game* management to the somewhat broader designation of *wildlife* management, a change reflected in the establishment and naming of the Wildlife Society in 1937, Leopold was already thinking more in terms of wildlife ecology. He looked forward to "an almost romantic expansion in professional responsibilities" in the wildlife field.

19. Elton concluded his *Ecology of Invasions* with a plea for the actual planning of a better and more varied landscape. "In no country has this been attempted with such remarkable drive and imagination," he observed, "as in the United States, where the spur for action has been soil erosion, combined with a fervent interest in preserving habitats for wild game" (p. 158). He did not specifically mention Leopold, but it is noteworthy that soil erosion and preservation of habitat for wild game were the spurs to Leopold's own action. Although *diversity*, as Leopold used the term toward the end of his life, implied coevolved diversity, Elton was willing to settle for *variety*, on the grounds that man's burgeoning population, his technology, and his penchant for mixing species the world over had made the preservation of coevolved diversity (wilderness) a moot question: "And provided the native species have their place, I see no reason why the reconstitution of communities to make them rich and interesting and stable should not include a careful selection of exotic forms, especially as many of these are in any case going to arrive in due course and occupy some niche" (p. 155).

Speaking on "The State of the Profession" in his presidential address to the Wildlife Society in 1940, Leopold observed that wildlife managers, who had begun with the job of producing something to shoot, might actually be contributing something far more important to the design for living. They might, without knowing it, be helping to write a new definition of the purpose of science. Most definitions of science dealt almost exclusively with the creation and exercise of power—"the idea of controlled environment," to use his own phrase of several years previous. "But," he was asking now, "what about the creation and exercise of wonder, of respect for workmanship in nature?" Shootable game was no longer very important to many "emancipated moderns," he pointed out, and not much game could be produced anyway until the landowner changed his ways of using land. The landowner in turn could not change his ways until his teachers, bankers, customers, editors, governors, and trespassers changed their ideas about what land was for. "To change ideas about what land is for," he mused, "is to change ideas about what anything is for."[20]

The new approach entailed not only a transmutation of values but also a renewed emphasis on broad public understanding. Deep-digging ecological research could invest wildlife with qualitative rather than merely quantitative value and, by revealing the drama of the land's workings, serve as a unifying force in a liberal education. Leopold's own undergraduate course at the university, Wildlife Ecology 118, which he began offering in 1939, was a highlight in the intellectual development of practically every student who was fortunate enough to stumble upon it. Its objective, as he explained, was "to teach the student to see the land, to understand what he sees, and enjoy what he understands."[21]

This was the period, especially during the early 1940s when World War II drew away nearly all of his graduate students, that Aldo Leopold wrote most of the literary

20. "The State of the Profession," *Journal of Wildlife Management*, 4:3 (July 1940), 343–46.
21. "The Role of Wildlife in a Liberal Education," *Transactions*, 7th North American Wildlife Conference (1942), p. 486.

and philosophical essays for which he is best known, among them "Great Possessions," "Odyssey," "Wildlife in American Culture," "Thinking Like a Mountain." It was a period during which he was involved in recommending new policy directions for approximately a hundred different professional societies and committees, conservation organizations, government agencies, research stations, conferences, and journals, culminating with his election in 1947 as president of the Ecological Society of America. And it was also in the 1940s that he reentered the realm of conservation politics, serving as a member of the Wisconsin Conservation Commission from 1943 until his death. The job took a tremendous toll on him, largely as a consequence of the leadership role he assumed in an effort to win public acceptance for a substantial reduction in Wisconsin's deer population, but it was his conviction that a man ought to expect to take on such responsibilities once in his lifetime.

Leopold's experiences in the public arena, particularly his efforts to bring about a reorientation in public thinking on the deer question in Wisconsin, reinforced his conviction of the need for an ecologically based ethic. Several times during the decade he struggled to express on paper his conception of an ecological ethic, and he finally succeeded some time in late 1947 or early 1948. Drawing from his "Conservation Ethic" of 1933 the notion of the cultural evolution of ethics and from his later "Biotic View of Land" the concept of the evolution of ecological diversity, and adding his strong conviction of individual responsibility for the health of the land, he produced his most important essay, "The Land Ethic." "A thing is right," he concluded, "when it tends to preserve the integrity, stability and beauty of the biotic community. It is wrong when it tends otherwise."[22]

These values, integrity (or coevolved diversity), stability, and beauty, were fundamental to Leopold's thinking from the beginning. But like his notion of a land ethic, they acquired new meanings and implications throughout his life in response to his changing perception of the environment, so they meant something quite different in the

22. *A Sand County Almanac and Sketches Here and There* (New York, 1949), pp. 224–25.

end from what they had in the beginning. The measure of this difference is in "Thinking Like a Mountain."

On April 21, 1948, Aldo Leopold died of a heart attack while helping his neighbors fight a grass fire that threatened his sand-country farm. One week earlier, the book of essays for which he had been seeking a publisher since early 1941 was accepted by Oxford University Press via long-distance telephone. It was published in 1949 as *A Sand County Almanac* with "The Land Ethic" as its capstone.

A Sand County Almanac represents the distillation of a lifetime of observation and reflection on the interrelations of ecology, esthetics, and ethics. Through it Aldo Leopold speaks to the present generation as he will to the future. The essays have a timeless quality, dealing as they do with ecological and evolutionary processes. Yet their strength comes from history, from Leopold's experiences in time, on the land.

Aldo Leopold's thinking was shaped by the land itself, and by his changing perception of it. He considered himself a field man. His thinking was not the product of books read or even of influential friends listened to, except as they made him think more deeply about what he saw in the land. It was his conviction that ecological perception was a matter of careful observation and critical thinking. It proceeded from a view of complexity to a sense of relatedness, a concern with causes and consequences. Leopold was not afraid to ask "Why?" but he did not attempt an answer seated at his desk. When one looks for critical junctures in his thinking, one finds them as often as not associated with some new field experience. He was extraordinarily willing to look and to see and to alter the contours of his thinking about a problem if what he saw warranted it. Yet he maintained a broad perspective on means and ends, grounded in the basic values of integrity, stability, and beauty, a perspective that enabled his ideas to grow and change naturally during the course of his life, and in the process impart greater depth, breadth, and clarity to his philosophy.

2

Southwestern Game Fields

Diversity and Dissolution

"The Southwest is a mosaic of the life and landscapes of North America," Aldo Leopold wrote in his unpublished book-length manuscript, "Southwestern Game Fields" (1927):

> You can throw on your pack in a Mexican jungle and in a few days climb to an Arctic tundra like that of Labrador. From your camp you can look back over Canadian forests of spruce and fir to California pineries and juniper foothills of Utah and Colorado. Through the twisted canyons which radiate from your feet run trout streams lined with Ozark hardwoods, debouching at the foot of the mountains upon honey-colored plains of the Panhandle. Across these plains are stretched green ribbons of Nebraskan rivers, which disappear on the horizon in the painted deserts of Arizona.

Each patch of the mosaic harbored its own characteristic animal species, from parrots and jaguars and javelinas to bighorn and ptarmigan and marmots, from ibises and egrets to trout and beaver and water ousels. "In the East," Leopold told delegates to a national game conference in New York, "you have one kind of deer—we have four, and antelope and mountain sheep and elk to boot. You have one kind of quail—we have three. You have one kind of grouse—we have three. You have one species of the family Columbidae, which includes doves and pigeons—we have three. And last of all, we have the lordly gobbler—there is no finer game anywhere."[1]

1. "Southwestern Game Fields," typescript, Ch. II, p. 1, 10 April 1927, LP 6B10; "The Game Situation in the Southwest," *Bulletin of the American Game Protective Association*, 9:2 (April 1920), 3. The East, of course, had the wild turkey until it killed off the population.

The diversity extended to human cultures as well; evidence of past occupation by cliff dwellers, Pueblos, Spaniards, and early Anglo-American settlers could still be read in the land. It was this diversity—diversity of altitude and topography, climate and water supply, vegetation, wildlife, and human cultures—that attracted artists, archaeologists, and historians, as well as Leopold himself. But "even among those who were born in the Southwest," Leopold observed, "not one in a thousand realizes what has happened to it,—that much of its beauty is the beauty not of life, but of dissolution."[2]

* * *

Leopold's romance with southwestern diversity, and probably also his consciousness of dissolution, began while he was on reconnaissance in the Blue Range and White Mountains of eastern Arizona, during those first years on the Apache Forest. The valley of Blue River was the probable route of Coronado on his quest in 1540 for the fabled Seven Cities of Cibola. Spaniards and Mexicans later grazed their sheep on the surrounding plains; but the mountainous plateaus and deep, rugged canyons along the Mogollon Rim remained the domain of hostile Apaches until Geronimo was captured in 1886 and the remnants of his tribe were confined to reservations. Then, into the lush bottoms of the Blue, stirrup-high in bunch grasses with groves of mixed hardwoods and pine and willow-lined trout streams, moved frontier ranchers from Texas, Oklahoma, and Kansas, seeking the last expanses of open range. Their wild longhorns wintered along the river bottoms and at lower elevations on the grass-covered pinyon–juniper foothills and summered higher up the slopes and 'on top', fattening on lush perennial grasses and herbs growing in broad, undulating alpine meadows and beneath perfectly open stands of magnificent ponderosa pine, fir, and aspen.

There was probably no country anywhere that had such a hold on Leopold. Years later, he reflected on it in prose:

2. "Southwestern Game Fields," Ch. II, p. 3.

The top of the mountain was a great meadow, half a day's ride across, but do not picture it as a single amphitheater of grass, hedged in by a wall of pines. The edges of that meadow were scrolled, curled, and crenulated with an infinity of bays and coves, points and stringers, peninsulas and parks, each one of which differed from all the rest. No man knew them all, and every day's ride offered a gambler's chance of finding a new one. I say 'new' because one often had the feeling, riding into some flower-spangled cove, that if anyone had ever been here before, he must of necessity have sung a song, or written a poem.

and in poetry:

> Have you topped the world at Salt-house
> With the mesas spread below you?
> A full hundred miles of daisies—
> Yellow islands in a sea—
> Dark blue deeps of threading canyons—
> Dim blue reach of far-off valley.
> Ho! the Salt House trail in autumn,—
> It is calling yet to me.[3]

It was the best game country in all Arizona. "In cruising I've seen everything from 12-point bucks and 30-pound gobblers to Mexican Pigeons and Wild Geese," Leopold wrote home to his sister, and he termed such sightings "hummocks on the mental prairie of the 'Reconnaisseur.'"[4] But even as early as 1909–1910 it would have been obvious to him that wildlife was nowhere near as abundant as it had once been on the Apache and that the very diversity of species was diminishing.

In the realm of big game, for example, the buffalo at which Coronado's men had marveled were gone, and the tropical peccaries, the jaguars, and mountain sheep were scarcely to be found. Merriam elk, which survived longer

3. "On Top," *A Sand County Almanac and Sketches Here and There* (New York, 1949), p. 126; a stanza from "Mesa de los Angeles," in John D. Guthrie, ed., *Forest Fire and Other Verse* (1929), p. 18. The mountain is the White Mountain; the Salt House trail ascends the Mogollon Rim in the Blue Range.
4. AL to Marie Leopold, 3 Oct. 1910, LP 8B4.

in the Blue than anywhere in the Southwest, had probably become extinct by 1900, although elk-rubbed trees and a few scattered antlers could still be spotted a decade later. Antelope were hanging on by a thread, protected by a congressionally closed season and the zealous efforts of forest officers. Wild turkeys were almost totally eliminated south of the Mogollon Rim. Of the deer, the whitetails that had dispersed in the brushfields seemed to be holding their own, but the blacktails (mule deer) which tended to congregate in the more open forests were on the wane, and the delicate Sonora deer (a sub-species of whitetail) was near extinction. Another class of wildlife, the predators, especially wolves and grizzlies, were being exterminated by local stockmen, government trappers, and forest officers, although no one including Leopold regretted their demise at the time. Some three decades later, however, in his essay "Escudilla," Leopold would lament the death by set-gun of the last grizzly, Old Bigfoot, robber-baron of Escudilla Mountain. The old wolf in "Thinking Like a Mountain" also may have been shot on the Apache Forest.[5]

As early as 1901, E. W. Nelson, later chief of the Biological Survey, had investigated the Blue Range–White Mountain area and reported on its desirability as a federal game preserve. The most serious menace to the game, he felt, were the Apache Indians in the nearby reservation, who characteristically set fire to the forest during spring and early summer when tabanid flies plagued the deer, in order to concentrate the animals in the sheltering smoke for easy slaughter. When Leopold arrived on the Apache in 1909 the elusive Indians were still "making jerky," and

5. Evidence on relative status of species in the early years is inconclusive and somewhat contradictory. This account is amalgamated from Leopold's own writings, the Apache newsletters, and E. W. Nelson, "The Black Mesa Forest Reserve of Arizona and its Availability as a Game Preserve," in Geo. B. Grinnell, ed., *American Big Game in its Haunts* (Boone and Crockett Club, 1904), 466–84. Leopold discusses deer species and status in "Southwestern Game Fields," Ch. III, pp. 2–9. As for Old Bigfoot, Leopold wrote to his father 25 Sept. 1909, LP 8B4, trying to entice him to come hunt on the Apache: "There is a big White Bear in here. Come on and show him how to behave. Nobody else can bother him so far, it seems."

he spent several days that fall searching for them, unsuc-
cessfully, and planned to devote a month to reporting on,
posting, and organizing a proposed Blue Range refuge,
though he never found time for the task. He and the other
forest officers on the Apache were appointed as assistant
fish and game commissioners of the Territory of Arizona
to help enforce the fish and game laws, such as they were.
The men with whom he was associated during those first
years on the Apache—Fred Winn, John D. Guthrie, Raymond Marsh, and others—were among the most earnest
advocates of game conservation anywhere in the Service;
but apprehending violators in so inaccessible a terrain
was no easy task and convictions were virtually impossible
to secure. That deer remained more abundant and other
species hung on longer on the Apache than elsewhere in
the Southwest was evidence of exceptionally favorable
environmental conditions in that area. If Leopold did not
appreciate this quality while he was on the Apache, he
must certainly have begun to be aware of it when he was
assigned to the relatively gameless Carson National
Forest.[6]

Game depletion was not the only evidence of dissolution
on the Apache. Leopold's Blue Range reconnaissance involved mapping and cruising timber along the route of a
proposed wagon road traversing the forest from Springerville south to the vicinity of Clifton, second-largest cop-
per-producing district in the nation. The pinyon–juniper
woodland to the south having been virtually denuded by
Mexicans supplying fuel to the mining towns, there was
now demand for the more inaccessible timber of the Blue
Range. The proposed logging road, however, would have
to clamber high over the mountains; it would be difficult
and expensive to construct and maintain, and it would be
open only four months of the year. There had once been
a more logical road following Coronado's route up the

6. Most of the early accounts of deer poaching on the Apache
Forest and elsewhere in the Southwest seem to have involved
Indians, perhaps because Indians were more successful in inaccessible terrain than white hunters or because they made more
appealing targets for white frontier justice. By the time he became involved in the game protective campaign, however, Leopold was more concerned about poaching by whites.

valley of the Blue, but it had been destroyed in January 1905 by severe flooding that had washed away nearly three-fourths of the once-lush bottomlands. It had been rebuilt that summer, only to be totally wiped out the following December by an even more severe flood, which completed the ruin of the agricultural land along the river and foreclosed all possibility of a road in the valley. Members of Leopold's crew recommended strongly against construction of the proposed road over the top and argued instead for dams, shear booms, flumes, and other stream improvements to permit driving logs down Blue River. Neither alternative was adopted and the timber remained standing, for the time being at least, but the conundrum must surely have given Leopold pause.[7]

From old settlers, he heard what the bottoms of the Blue had been like before the floods. He must have been aware of the difficulties the few doggedly persisting ranchers were experiencing without the land on which they had depended for alfalfa and grain fields, garden patches and orchards to sustain their livestock operations and their families in so isolated a valley. He listened to cowmen complaining that a shortage of grass farther up the slopes reduced greatly the number of cattle that could be supported and that recent incursions of dense brush made it nearly impossible even to find their cattle in places where in the 1880s they might have spotted a steer miles off. How Leopold interpreted these changes at the time we do not know.

By the time he revisited the Apache on an inspection trip in 1921, he knew of nearly two dozen other mountain valleys in the national forests of the Southwest that were

7. The account of the floods is drawn from W. H. Kent, "Clifton Addition to the Black Mesa Forest Reserve," typescript, c. 1906, 14 pp., National Archives, Record Group 95 (hereafter cited as NA–RG95), Research Compilation File. The report on the road is in a letter, Daniel W. Adams to District Forester, 6 Oct. 1909, and related documents in the historical files of the Apache National Forest, Springerville, Arizona. Leopold wrote to his mother 7 Oct. 1909, LP 8B4, indicating his eagerness to handle the big "15-million a year" timber sales when the time came, but commenting "I have some figures on this fool mountain road that will make [the assistant district forester] squirm."

already wholly or partly ruined or were in the process of being eroded. It seemed obvious to him by this time that the principal cause of accelerated erosion was overgrazing in the interval since white settlement. On the Apache, meanwhile, the automobile had become a factor and with it the "good roads mania," and the government was now prepared to spend half a million dollars on the road over the Blue Range that had seemed so preposterous a decade earlier. The new road could not tap the remaining home-steads along Blue River, yet it was to be designated, iron-ically, the Coronado Trail. In a speech to the New Mexico Association for Science, Leopold wryly summed up his misgivings about the economics and the logic of the commonly accepted approach toward settling and recla-mation of the Southwest, using the Blue as his example:

> We, the community, have "developed" Blue River by
> overgrazing the range, washing out half-a-million in land,
> taking the profits out of the livestock industry, cutting the
> ranch homes by two-thirds, destroying conditions necessary
> for keeping families in the other third, leaving the timber
> without an outlet to the place where it is needed, and now
> we are spending half-a-million to build a road around this
> place of desolation which we have created. And to replace this
> smiling valley which nature gave us free, we are spending
> another half-a-million to reclaim an equal acreage of desert
> in some place where we do not need it nearly as badly
> nor can use it nearly so well. This, fellow citizens, is Nordic
> genius for reducing to possession the wilderness.[8]

In a fitting capstone to this bit of history, the Forest Ser-vice in 1933 added the Blue to the national forest wilder-ness system.[9]

8. Erosion as a Menace to the Social and Economic Future of the Southwest," *Journal of Forestry*, 44:9 (Sept. 1946), 629. Al-though this speech, delivered in 1922, was not published until 1946, Leopold used the Blue as an example of erosion in several other articles during 1921–1924.

9. Under Regulation L–20, the Chief of the Forest Service designated about 200,000 acres of the Blue east of the Coronado Trail as the Blue Range Primitive Area in 1933. Under terms of the Wilderness Act of 1964, which provided for further study of

The Virgin Southwest and What the
White Man Has Done to It

The diversity and dissolution of the Blue led Leopold to question whether white Americans were capable of maintaining a viable civilization in the Southwest. As he mused on "The Virgin Southwest and What the White Man Has Done to It" (a chapter in "Southwestern Game Fields"), he became conscious of one basic fact written across the face of the land:

> that this very diversity of climate and water supply which made these hills and plains a treasure house of wealth and beauty likewise made them a fragile dwelling for the modern white man. The engines wherewith he conquers these rocks and rills and templed hills are stronger than his understanding of what hills are, and more powerful than his vaunted love for them. . . . The hair-trigger equilibrium of natural forces, but for which Coronado and Zebulon Pike would have discovered a desert, has been upset, which was inevitable, and then ignored, which was not.
> (pp. 2–3)

Fascinated by the accounts of explorers and settlers and with his own uncanny skill at reading history backward in the land, Leopold by the early 1920s had formulated a stunning explanation of the forces at work in maintaining and upsetting the hair-trigger equilibrium of southwestern watersheds. It was an explanation remarkably in tune with modern ecological thinking, except perhaps in his dismissal of the factor of climatic change. Realizing that the question of climatic change was crucial in determining whether he was dealing with an "act of God" or merely with the consequences of unwise use by man, he

---

primitive areas to determine their suitability or nonsuitability for inclusion in the National Wilderness Preservation System, the Forest Service in 1971 proposed a Blue Range Wilderness of 177,239 acres for public consideration. As of June 1974 it was stalled in Congress because of opposition from mining interests, including Phelps-Dodge Company, which was drilling for copper in the heart of the area.

had examined the available evidence with care. Having satisfied himself that the climate had been generally stable during the last three thousand years, except of course for periodic fluctuations to which man simply had to learn to adjust, he went on to consider other factors in the southwestern environment.[10] One can hardly do justice to his achievement without analyzing either the process by which Leopold arrived at his conclusions or his understanding of the mechanism of erosion, the prime agency of dissolution. But we can at least sketch the outlines of his interpretation of environmental change, as a basis for the discussion of wildlife conservation and deer management that follows.

Leopold made his most penetrating observations on environmental change in the first region in which he worked, "that tumbled sea of pale blue hills that beats against the foot of the mighty escarpment known as the Mogollon Rim," stretching from the Prescott and Tonto national forests of central Arizona two hundred miles southeastward through the Blue Range to the Gila National Forest in New Mexico. "Spread upon these hills," wrote Leopold, "are brushfields so vast that the eye can barely see the yellow deserts that lie beyond, and so dense that the cowmen must wear 'Blue River Chaps'—great double aprons of thick leather covering not only the rider's legs but the horse's flanks as well, all polished a shiny bronze by the constant rubbing of oak and buckbrush, manzanita and mesquite." These hills, like the ponderosa pine forests on top and the bottomlands in the valleys, were once bound-

10. See "Some Fundamentals of Conservation in the Southwest," typescript, c. 1923, LP 6B16. Since 1923, tree-ring analysis has become more sophisticated and carbon-14 techniques have aided in the dating of archaeological evidence, so evidence that climatic change occurred is more definite; but there is still a lively debate concerning the nature of climatic change and its relationship to other environmental changes. See, for example, the work of Leopold's son, Luna B. Leopold, "The Erosion Problem of Southwestern United States" (Ph.D. dissertation, Harvard, 1950), some chapters of which were published as articles; and a review of various interpretations in James Rodney Hastings and Raymond M. Turner, *The Changing Mile: An Ecological Study of Vegetation Change With Time in the Lower Mile of an Arid and Semiarid Region* (Tucson, 1965).

ed by a continuous sea of grass, such as greeted the early settlers on the Blue.[11]

From gnarled and ancient junipers—those "chronographs of ecological revolution," he called them—Leopold learned one of the key factors in the virgin equilibrium: fire. Healed fire scars in the living wood of junipers five hundred years old attested to centuries of periodic fire, occurring every ten years or so, he thought, but ending about forty years previous, just after white settlement. The fires, set by lightning or by Indians trying to concentrate the deer for easy hunting, kept the brush thinned out and gave a competitive advantage to the grass and to trees that had managed to become established in the intervals between fires. With the introduction of hundreds of thousands of cattle and sheep in the 1880s, however, overgrazing rapidly thinned out the grass, which had been the connecting medium necessary to carry fires; and the brush, freed of root competition from grass and no longer held in check by periodic fire, began to "take the country" under the Mogollon Rim. Higher up on the plateaus, yellow pine that had seeded in on soil exposed by overgrazing or by the latest fires, began to grow up in dense thickets, instead of the open, grassy stands the explorers had described. By the time the brush and the pine had grown dense enough to carry fire, the Forest Service had arrived on the scene, with its policy and techniques of absolute fire prevention.[12]

In his observations of environmental processes Leopold

11. "Southwestern Game Fields," Ch. II, p. 21. Although the discussion that follows is based largely on the chapter, "The Virgin Southwest and What the White Man Has Done to It," it also draws from Leopold's earlier published analysis of "Grass, Brush, Timber, and Fire in Southern Arizona," *Journal of Forestry*, 22:6 (Oct. 1924), 1–10; from a mimeographed "Watershed Handbook" he prepared for District 3 (Dec. 1923); and from a number of his other published and unpublished writings from the period.

12. Since Leopold was describing the Mogollon country, which had been the domain of the Apaches until the mid-1880s, he could tie the beginnings of overgrazing to the advent of American settlers. Elsewhere in the Southwest, however, he recognized that Spaniards and Indians had overgrazed their lands before Anglo-Americans had arrived.

was an ecologist, advanced beyond most other ecologists of his day in recognizing the role of periodic disturbance by fire, while in his evaluation of the changes and in his prescriptions for management he was very much the traditional forester. As ecologist he observed pine and juniper reproduction spreading into formerly open park and meadow areas, and also that the boundaries of various types of vegetation were moving downhill; pine was spreading into the juniper-and-brush of the formerly grassy woodland zone, and juniper was beginning to reproduce in the thick brush of the semidesert type. As a forester and a game conservationist, Leopold did not at all regret "that cattle have thus unwittingly restored to the forest its lost provinces"; but as a citizen concerned with the long-term prospects for civilization in the Southwest, he was appalled at the widespread erosion that was accompanying this transition. His theory about the ecology of the brushfields, he pointed out, involved a recognition that grass conserves watersheds much more effectively than brush and that overgrazing, not fire, was the factor most destructive to watershed values, similarly in the forest zone. Despite his forester's commitment to timber production, Leopold the ecologist challenged the traditional doctrine of 'forest influences'—the belief that forest trees high up the slopes at the headwaters were the most crucial factor in preventing destructive floods downstream. It was grass cover and the interlocking mesh of grass roots that was most crucial in maintaining the stability of watersheds, he believed, and it was earth scars caused by localized overgrazing and trampling that gave destructive erosion and flooding its start.

Overgrazing, in the rough topography of the Mogollon country, was so insidious because it could occur in localized areas long before all available grass or browse had been eaten. Hence, Leopold drew the conclusion that the number of cattle permitted on national forest ranges ought to be determined by the condition of the watersheds, with the virgin condition as the standard, rather than by the usual Forest Service criteria of whether the available forage could support more cattle and whether heavier grazing might appreciably reduce the fire hazard. To the fire-hazard justification for heavy grazing, he countered that damage to watersheds was far more costly than the potential damage of fire or the added cost of fire protection, in the brushfields at least. The fire hazard of brush could hardly be reduced by grazing anyway, he pointed out, because grazing pressure on palatable species would simply result in their replacement by unpalatable species.

Although he was willing to take a slight added risk of fire in exchange for more conservative grazing, he was too much imbued with the traditional Forest Service doctrine of absolute exclusion of fire to suggest that fire actually be utilized as a tool for managing land. Controlled burning could have been used to thin out the brush, thereby promoting restoration of the grass cover; and in the pine forests, fire could have been utilized to burn accumulated litter in order to reduce the hazard of more serious 'crown fires'. Years of observation in the Southwest, however, had convinced Leopold that, although fire could admittedly help to destroy brush and accumulated litter without killing mature trees, even the so-called light-burning or "Piute forestry" favored by the Southern Pacific Railroad and certain timber interests was in fact a serious drain on forest productivity. Not only did fire destroy soil humus necessary for rapid growth of trees and inflict scars that increased the susceptibility of trees to disease and insects, but even more serious to his mind, it destroyed the young seedlings that were the future forest. He was, moreover, convinced that fire was one of the greatest enemies to game, killing not only directly and by destruction of coverts but also by gradually eliminating the forage plants most useful to game.[13]

13. Leopold's views on fire and forests are expressed most directly in " 'Piute Forestry' vs. Forest Fire Prevention," *Southwestern Magazine*, 2:3 (March 1920), 12–13. He seems to have thought that light-burning propagandists were interested only in protecting their investments in mature timber, in some cases for speculative profit, and were not at all concerned about young growth and new reproduction. His views on wildlife and fire may be found in "Wild Followers of the Forest," *American For-*

A return to the days of periodic fire, whether wild or controlled, was thus no answer to the erosion problem for Leopold. Complete exclusion of grazing by cattle and sheep was no answer either, for these animals enabled forest reproduction to spread by thinning out the native grasses and, in the case of sheep, by depositing seeds in their droppings. The challenge for Forest Service research and administration, as he posed it, was to devise artificial works for erosion control and to so regulate grazing and

---

estry, 29:357 (Sept. 1923), 515–19, 568. In some regions, Leopold admitted, food plants useful for game increased in the aftermath of fire, as in the north woods where berries valuable for bear and grouse became abundant. "For this exception to an otherwise black record," he wrote, "let the fire devil have his due." He then went on to point out that the practice of forestry, which is to say the cutting of mature timber, likewise opened up the forest canopy and allowed sun-loving food plants to reproduce, without the enormous damage to game and forests entailed by fire.

Perhaps because of the official resistance it has encountered in the U.S. Forest Service, the idea of controlled burning, like the idea of absolute prevention of fire, has become somewhat of an evangelistic cause. For a discussion of the controversy see Ashley L. Schiff, *Fire and Water: Scientific Heresy in the Forest Service* (Cambridge, Mass., 1962). Leopold, although he was a pioneer in the field of fire ecology, never became a controlled-burning enthusiast. In later years he recognized the potential for maintaining prairies and certain successional stages in marshes by burning, but as to burning on forest lands he seems to have reserved judgment.

The evidence on fire ecology is far from complete, although the atmosphere is gradually becoming more conducive to broad-based, critical research. Controlled burning is now accepted practice in southeastern pine forests, and the Forest Service has begun experimenting with fire elsewhere, including yellow pine forests and pinyon—juniper woodland in the Southwest. Ecologists are now becoming interested in the use of fire in the preservation of wildlands. See, for example, A. Starker Leopold, et al., "Wildlife Management in the National Parks," 4 March 1963, and "The National Wildlife Refuge System," 11 March 1968, both of which are reports of the Secretary of the Interior's Advisory Board on Wildlife Management; Estella B. Leopold, "Ecological Requirements of the Wilderness Act," in Maxine E. McCloskey and James P. Gilligan, eds., *Wilderness and the Quality of Life* (San Francisco, 1969), pp. 188–97; and the *Proceedings* of the Annual Tall Timbers Fire Ecology Conferences.

other land-use practices as to conserve the benefit to timber while maintaining the stability of the watersheds.

Not one to regard any aspect of the natural world as
static, Leopold conceived of virgin watersheds as having
been maintained in a dynamic equilibrium of erosive
and resistant (or restorative) forces. At any given time
some areas would have been eroding and others alluviating, but the net result over, say, a generation was a landscape of roughly the same configuration. It was this long-
term stability that Leopold valued. Yet he did not regard
the virgin grass cover as a climax condition or as necessarily most desirable; the implication is clear that he was
delighted to witness the return of the forest to areas from
which it had been "expropriated by centuries of fire."
There were many possible levels of equilibrium, or stability, he seemed to think, running all the way from the
unproductive mechanical equilibrium of cobbles in Blue
Valley to the fire-maintained grass cover of presettlement
days to a new climax equilibrium of thick pine forest,
which he apparently hoped would be highly productive
of both timber and game.

It is ironic that the very features of the southwestern
landscape that were most attractive to Leopold—the
grass-covered pinyon–juniper foothills, the open sunlit
stands of ponderosa pine, the "infinity of bays and coves,
points and stringers, peninsulas and parks," the endless
vistas—in short the very openness and diversity of the
country, were even then trending toward a monotony of
brush and pine too impenetrable for a man to move about
in freely, too dense to support the variety of grass, herbs,
and browse required for diverse and abundant wildlife,
and too thick even for proper growth and health of the
trees themselves. Leopold as ecologist observed the beginnings of this process; as forester he rejoiced in new thickets
of sapling pine and the spread of juniper into formerly
treeless meadows; as game conservationist he thrilled to
the sight of whitetails fattening on the abundant new
growth of brush on the foothills; but he did not, *could
not*, imagine the extent to which this process would be
carried under the regimen of grazing and exclusion of fire.
Nor did he have any basis at the time for appreciating

the detrimental effect it would ultimately have on both timber and wildlife, as well as on livestock.[14]

Valuing diversity as he did, it was natural that he should ponder its relationship to the processes he was observing—especially the process of erosion, which was his greatest concern at the time. He sensed that diversity was somehow related to the stability of the environment; yet as sometimes happens when one is groping for a relationship not yet fully comprehended, he fastened at first on an explanation that was seemingly the reverse of the position he was later to maintain. As he expressed it in "Southwestern Game Fields," diversity was somehow the cause of the erosion problem, or instability:

> the very diversity of types which characterized the Southwest has been a primary cause of its present despoliation. If a mountain cow on a cold winter day has the choice of basking in the warm sun of a Missouri woodlot, or climbing up into the wind-swept Panhandle, or scrambling among the rocky slopes of Colorado, she will choose Missouri. In fact she may browse the last Missouri willow to death before the Colorado bunch grass is even touched. But the erosion thus induced on the creek bottom has no compunctions about spreading to adjacent territory, be it overgrazed or not. The greater the diversity of types, the less uniform their utilization and the quicker will a given amount of grazing induce erosion.

This passage contains an important insight. In an arid environment, organic resources bear delicately balanced

---

14. An excellent analysis of environmental change in the pine forests along the Mogollon Rim, especially in the vicinity of the Apache National Forest, is Charles F. Cooper, "Changes in Vegetation, Structure, and Growth of Southwestern Pine Forests Since White Settlement," *Ecological Monographs*, 30:2 (April 1960), 129–64. For changes in the pinyon–juniper zone see Joseph F. Arnold, et al., "The Pinyon-Juniper Type of Arizona: Effects of Grazing, Fire and Tree Control," Production Research Report No. 84, USDA—Forest Service, Sept. 1964. Wildlife habitat relations in the various types are discussed in a number of research notes of the Rocky Mountain Forest and Range Experiment Station, and other papers, notably by Hudson G. Reynolds.

relationships to each other and these relationships frequently extend across type boundaries, so that any upsetting of the balance in one area could cause progressive deterioration in other areas. Yet it was not the diversity of types that caused erosion but the diminishing of diversity within types by overgrazing. Years later, with a firmer grasp of functional ecology, Leopold would conceptualize stability of the land mechanism as a function of ecological diversity and instability as a function of violence in land use, which usually caused a reduction in diversity.[15]

His fascination with the concept of diversity, despite his lack of a firm ecological understanding of it, is further evidenced in his attempt to interpret the effect of environmental changes on wildlife. Leaning over backwards to be fair to the white man, he presented a positive aspect of the Blue River erosion story. The silt from erosion in the mountains together with the irrigation works required to replace agricultural lands lost to that erosion seemed to be resulting in an increase of diversity in the desert:

> Have you ever seen blue quail darting wraithlike through
> the willows of a ditch-bank? Listen to their whistling soft
> as November sunlight, and learn that these are the willows
> that once anchored a stream bank in some mountain val-
> ley. Trout hid under their roots, and in winter deer browsed
> on their tender twigs. Now in these fields that once were
> desert these same willows anchor this same soil to winding

15. "Southwestern Game Fields," Ch. II, pp. 12–13. In an earlier manuscript, "Some Fundamentals of Conservation in the Southwest," typescript, c. 1923, 19 pp., LP 6B16, Leopold had viewed stability as related to climate: "There seems to be a natural law which governs the resistance of nature to human abuse. Broadly speaking, the law is this: the degree of stability varies inversely to the aridity." He was later to discover, in the course of supervising work on erosion control in the southwestern national forests in 1933, that some of the best plants for revegetating and restoring eroded watercourses were species that were no longer found on most watersheds because they had been grazed out by cattle before the start of destructive erosion —evidence, perhaps, that violent land use in the form of overgrazing by cattle, especially in an arid climate, had resulted in a diminishing of diversity and a consequent weakening of the vegetative resistance to erosion.

irrigation ditches, and cover it with long ribbons of ochre-
colored thickets for quail to play in.

In short, the diversity of mountain life and landscape
which erosion has torn away is here resurrected in what
once was desert.[16]

Leopold was beginning to recognize the complexities
of environmental change and he was fascinated by them;
yet he found it virtually impossible to evaluate their
over-all effect on game. The encroachment of brushfields,
for example, seemed to be a boon to black bears and also
to white-tailed deer, although mule deer were decreasing.
Reproduction of pine and juniper in formerly open
areas, he noted, provided additional coverts for deer but
deprived them of places to escape from flies. Wild turkeys
had been exterminated from most of the ranges taken by
brush, whereas they had merely been decimated elsewhere. Mountain sheep had dwindled, but who knew the
reasons why? Erosion had torn out streamside vegetation
required by most species of wildlife, but agriculture and
livestock developments in other places provided alternatives. Leopold granted that the net effect of manmade
changes probably came "nearer being beneficial" to game
in the Mogollon country than in any other southwestern
region. But beyond that he would not venture:

> We cannot evaluate these changes and strike off a bal-
> ance. We are in the position of a biographer who cannot
> evaluate a contemporary because he has too many facts
> and too little understanding of what they mean. Moreover
> the forces at work on our subject are still in operation;
> it is too early to foresee their final effect.

16. "Southwestern Game Fields," Ch. II, pp. 15–16. Leopold
noted that ducks and geese, cranes, beavers, and prairie chickens, like quail, had all extended their ranges in the Southwest
with the extension of agriculture and irrigation. Elsewhere in
the same chapter, however, he explained how siltation from
the mountain valleys was desposited in the channels of the main
rivers, like the Rio Grande, eventually raising the river channels
and irrigation ditches higher than the valley floor, so the river
"bogs its own bottoms with seepage and poisons their fertility
with alkali."

"This land," he concluded, "is too complex for its inhabitants to understand; maybe too complex for any competitive economic system to develop successfully. For the white man to live in real harmony with it seems to require either a degree of public regulation he will not tolerate, or a degree of private enlightenment he does not possess. But of course," he added, "we must continue to live with it according to our lights." He saw two possibilities for improving those lights. One was to apply science to land use. The other was to cultivate that love of country based on respect for the living earth, which he would later term a *land ethic*.[17]

The conservation of game by management, in this context, was conceived by Leopold as one small test of man's ability to apply science to land use and to maintain a permanent functional civilization in harmony with his natural environment. It was the deer, as indicated earlier, that Leopold in these years regarded as the numenon of the southwestern mountains. "Even from the open plains the mountain skyline derives a flavor from the thought that in its hazy labyrinths of ridge and canyon there are deer. What is the outlook for keeping them there?" he queried. "What can an intelligent self-directive civilization do to use and enjoy this country without destroying its savor?"[18]

This was the context in which Aldo Leopold presented the problem of deer management in "Southwestern Game Fields" in the mid-1920s. Before going on to analyze his approach to a science of deer management and the challenge to that approach posed by the succession of events out in the game fields, we shall sketch his activities as a game protectionist during about 1915–1920 and the early development of his ideas about forestry and game management.

Game Protection: The Cause

Aldo Leopold's early activities in game protection represent a telescoped version of a five-step sequence he was

17. Ibid., pp. 27–28.
18. "Southwestern Game Fields," Ch. III, pp. 1–2.

later to outline in *Game Management*. It was a sequence that he thought could usually be identified in the history of ideas about game conservation, whether in Europe, Asia, or America:

1. Restriction of hunting.
2. Predator control.
3. Reservation of game lands (as parks, forests, refuges, etc.).
4. Artificial replenishment (restocking and game farming).
5. Environmental controls (control of food, cover, special factors, and disease).[19]

His earliest efforts on the Apache and the "Game and Fish Handbook" he prepared for District 3 in 1915 were geared to enforcement of restrictions on hunting. As he stumped New Mexico and Arizona organizing game protective associations he carried the gospel of predator control and the Hornaday plan for wildlife refuges. Through the newly organized associations he promoted artificial replenishment of depleted game lands and trout streams. Within a few years he would be pointing the way to environmental controls.

Roughly parallel with this sequence of techniques, Leopold discerned an even more important evolution of objectives. Thus game laws, predator control, and reservation of game lands were originally expressions of a restrictive objective; they were devices for stringing out or dividing up "a dwindling treasure which nature, rather than man, had produced." The emphasis on restriction of kill had already yielded in some quarters to the objective of production or cropping of game, which at first meant raising animals on game farms or artificially replenishing lands with transplanted exotics. The significant breakthrough, to Leopold's mind, came in the first

19. *Game Management* (New York, 1933), pp. 4–5, 17. Although in outlining the sequence of techniques Leopold indicated that environmental controls dealt with food, cover, special factors, and disease, it is clear elsewhere in *Game Management* and in other writings that he regarded environmental control as roughly synonymous with encouraging production in the wild, as opposed to artificial propagation. In this larger sense, environmental controls included control of hunting, predators, and refuges as well.

decade of the twentieth century with Theodore Roosevelt and his objective of "conservation through wise use." The conservation idea, as applied initially to forests, involved a conception of timber as a renewable organic resource, capable of being produced by essentially natural means and harvested on a sustained-yield basis. Leopold was among the first to apply the idea to game, to conceive of producing native species in the wild and harvesting the surplus on a sustained basis. Yet despite his optimistic commitment to the production of game, he made his mark in the early years fighting the classic battles for restriction of kill.

Leopold seems to have been realistically content to preserve mere remnants of most big-game species that were already perilously near extinction, such as mountain sheep and antelope, with no thought of producing a surplus for hunting. With regard to these species, then, both his objective and his technique were essentially restrictive or "protectionist." Deer were the only big game left in the Southwest in anywhere near sufficient numbers for hunting, so to get across the new conservation idea Leopold concentrated right from the start on deer.[20] Yet the techniques he advocated for producing deer in the early years were essentially those that involved restricting the kill—game laws, predator control, and refuges. While the history of ideas in game conservation may have involved a discernible sequence of techniques and objectives in other countries or for other people in the United States, for Leopold the sequence was compressed to a stage.

Maintaining shootable deer seemed formidable enough. Reports submitted by forest officers in New Mexico under a census procedure initiated by Leopold showed a total of only 656 deer killed by hunters on 13,000 square miles

20. Of perhaps 20,000 mountain sheep "originally" in New Mexico, for example, Leopold estimated there were no more than 150 individuals remaining; of 100,000 antelope, he estimated 1,740; and of 200,000 deer, probably 10,000 left. "What has Posterity Done to Us?" holograph, c. 1916, 1 p., LP 4B1. Implicit in his argument for preservation of diversity was the supposition that deer hunters would be more attracted to an area in which they had a chance of at least seeing some of the other species.

of national forest land in New Mexico during the season of 1915. This was only half the number killed in 1914, and many foresters considered a five-year closed season essential to save the deer from virtual extermination. Leopold, unlike the classic protectionists such as William T. Hornaday, was very much opposed to complete closed seasons if there were any viable alternatives. Game became too tame without hunting seasons, he explained, with resulting slaughter when the season was reopened; closed seasons were difficult to enforce against local residents; and not least, they were "a negation of the fundamental idea of *use,* which underlies the administration of the national forests." Part of his argument for game conservation was premised on the economic value of big-game hunting, which of course would be eliminated were the season closed on all species. He calculated the value of deer killed in New Mexico in 1915, for example, at $43,296.00, figuring 15¢ a pound for the meat, $1.00 for the raw hides, and about $50 in recreational expenditures (transportation, provisions, equipment, license, and $5 per day for 5 days in time) for each successful hunter. Adding the expenditures of unsuccessful hunters would nearly double the value. An economic rationale was no less important in 1915 than in a later day.[21]

To combat the idea of closed seasons and win support for his concept of conservation, Leopold had to demonstrate the potential for vastly greater production of deer. He pointed to the examples of "thickly settled" eastern states like New York, Michigan, Maine, and Vermont, which sustained an annual kill of roughly one deer per five square miles. The 1915 figures for the comparative wilderness of the New Mexico national forests worked

21. "656 Deer Killed in New Mexico Forests," typescript, 26 Jan. 1916, 1 p. (press release, Forest Service, Albuquerque Office), LP 4B1; "Wanted—National Forest Game Refuges," *Bulletin of the American Game Protective Association,* 9:1 (Jan. 1920), 10; "Value of Big Game Killed, National Forests of New Mexico, 1915–1916," compiled by AL, 1 Oct. 1917, NA–RG95, Records of the Division of Wildlife Management. On the million-acre Carson Forest, of which Leopold had been supervisor 1911–1913, only eight deer were reported killed in 1915.

out to one deer per twenty square miles of range, only one-fourth the productiveness of the civilized East. "Who shall say," queried Leopold, "that only a wilderness can raise deer?"[22]

The common view that the advance of civilization spelled the doom of wildlife Leopold termed *the wilderness fallacy*. The same fallacy had once characterized our attitude toward forests:

> When the pioneer hewed a path for progress through the American wilderness, there was bred into the American people the idea that civilization and forests were two mutually exclusive propositions. Development and forest destruction went hand in hand; we therefore adopted the fallacy that they were synonymous. A stump was our symbol of progress.

Years of moral fervor in the cause of forestry, backed up by scientific research and sound policy in the management of the national forests, were finally beginning to persuade the American people that forests were not only compatible with civilization but essential to it. Leopold looked forward to the same transformation in attitudes toward wildlife: "Progress is no longer an excuse for the destruction of our native animals and birds," he observed, "but on the contrary implies not only an obligation, but an opportunity for their perpetuation."[23]

Of all big-game animals, deer were probably the most compatible with civilization. Perhaps that is why they alone maintained shootable populations in the Southwest. Leopold conceded that disappearance of the huge migratory herds of buffalo was inevitable, and he was extremely wary of restocking any but the most remote reaches of the southwestern mountains with elk, which competed with livestock for forage and caused serious

22. "The Popular Wilderness Fallacy," *Outer's Book-Recreation*, 58:1 (Jan. 1918), 46. In an earlier article, "Game Conservation: A Warning, also an Opportunity," 7:1-2 (1916), 6, Leopold had indicated that the New England states produced a deer per two square miles, for a preponderance in productiveness of 10:1 over the Southwest.

23. "The Popular Wilderness Fallacy," pp. 43, 46.

depredations around ranches. Years later he would argue that wilderness was essential for preservation of certain predators like grizzlies and wolves, which in the early years he thought only in terms of eradicating. But he could see no conflict whatever between deer and civilization, at least not in the national forests of the Southwest. Deer, he pointed out, subsisted mainly on browse (tender shoots, twigs, leaves, and fruits of shrubs and trees), while cattle fed largely on grass. "Generally speaking, there is no stock range not actually overgrazed," he asserted, "which does not have enough excess or inaccessible feed to support all the game any reasonable man would want to see."[24] It seems never to have occurred to him in those days that deer, like elk and cattle, might also "overgraze" a range, nor that deer, like sheep, might inhibit forest reproduction.

Not only did deer seem not to be in conflict with livestock and forestry operations—the "civilization" of the southwestern mountains—but they seemed actually to be abetted by provision of artificial reservoirs and salt licks for livestock, campaigns against predatory animals, and the virtual exclusion of forest fires. Overgrazing by livestock had undoubtedly been inimical to deer, but in his far-reaching optimism Leopold saw even the end of overgrazing and the recovery of the ranges. It should be noted that in his early years as a game protectionist and perhaps later, he thought more in terms of presence or absence of food and cover for game than of the effects of qualitative change in vegetation. Since the economic activities of civilized man seemed to be taking care of food, water, and cover, what was needed now were means to preserve a breeding stock of deer sufficient to repopulate these increasingly habitable ranges. There were three means: law enforcement, predator control, and refuges.

"Why is the big game of the West disappearing today? Principally for the reason that the game laws are not enforced. And why are they not enforced? *Politics.* Paper game laws and political game wardens are one and inseparable." Thus began an article in which Leopold told

24. "Wanted—National Forest Game Refuges," p. 9.

how New Mexico sportsmen, through their newly organized game protective associations, laid strategy during the 1916 gubernatorial campaign to secure the appointment of an "efficiency game warden" whom they could respect. "Sportsmen do not *really support* a game warden," he observed, "until they have had to *roll up their sleeves* in some common disinterested effort to get a good one." A new day had dawned for enforcement of game laws in New Mexico—that is, until the next governor took office in 1919 and declined to reappoint the sportsmen's efficiency warden. In a stinging rebuke, "Wants New Mexico Soviet," the *Albuquerque Morning Journal* accused Leopold of having been "offensively active" in his political stratagems aimed at controlling the appointment of the warden. For the game protective associations to have the right to name or approve the warden as they demanded (on the grounds that the sportsmen provided financial support for the state game department) would be soviet government, government by special interests, the *Journal* pointed out.[25]

Whether because they lost out in 1919 or because they recognized the kernel of truth in the *Journal's* invective, Leopold and his fellow sportsman–activists took a new tack, drawing up and promoting a legislative proposal for a "nonpolitical" state game commission, which was enacted by the New Mexico Legislature in 1921. The new law provided for a three-man, unpaid commission authorized not only to enforce game laws but also to encourage production of game. However, the commission was not empowered to liberalize any seasons or bag limits set by the legislature or to declare a closed season on any predatory species; in New Mexico in 1921 it was inconceivable that it would ever be desirable for the commission to have such authority.

Law enforcement, no matter how strict, could not save the meager remnant of deer, not when predatory animals

25. "Putting the 'AM' in Game Warden," *The Sportsmen's Review* (undated clipping, c. 1917), LP 6B1; "Wants New Mexico Soviet," *Albuquerque Morning Journal* (undated clipping, c. July 1919), LP 8B2.

were free to hunt 365 days of the year. Hence the rationale for the sportsmen's war on predators. Bounties on wolves and other predators on domestic livestock were paid from earliest settlement of the country, and by the 1900s wolves were effectively eradicated from much of the East. In the relative wilderness of the western states, however, where the wanton slaughter of buffalo, elk, antelope, and other big-game animals threw the full weight of predation on livestock and on the few remaining deer, the combined efforts of bounty hunters, stockmen, and forest officers were unable to control the depredations, valued at millions of dollars annually. Consequently, Congress in 1915 appropriated funds for the U.S. Biological Survey to start an intensive campaign against predators, employing a force of more than three hundred hunters. The work in New Mexico and Arizona was under the direction of J. Stokley Ligon, who was to become a personal friend of Leopold's and a coauthor of "Southwestern Game Fields."

In New Mexico, the coalition of sportsmen and stockmen forged by Leopold secured establishment of a State Council of Defense, funded by the legislature, which pooled its money with the Biological Survey to make that state one of the leaders in predator eradication. By 1920 Leopold could report to the National Game Conference that about 300 wolves in the state had been reduced to an estimated 30 in only three years. There were still about 100 adult mountain lions left in New Mexico, however, each one killing at least 30 deer per year, while hunters were taking only about 700 deer a year. Thus lions alone, he asserted were more than four times as destructive as hunters. (Mountain lions actually killed more deer than did wolves and were seldom found except in deer country; but wolves, feared throughout history as killers of domesticated animals and even of people, had become both symbol and scapegoat of the predatory species and were thus more zealously eradicated.) As the eradication campaign progressed, the remaining animals were going to become more difficult and more expensive to catch, Leopold warned; it was thus imperative that the organized sportsmen throw their full weight behind the Biological

Survey and insist that the job be completely finished—to the very last animal, he emphasized—lest in a few years wolves and lions again reclaim the ranges.[26]

It is natural to question, in view of the significance Leopold attached to predation in his later thinking, whether his frequent exhortations to complete eradication of big-game predators, which he termed *varmints* or *vermin*, represented a carefully thought out philosophical position or merely the unfortunate overstatements of an enthusiast. There is no satisfactory answer. In any case, never during 1915–1920 did he specifically include any of the species of big predators in his pleas for preservation of native wildlife. He possibly did not think of *varmints* as wildlife, much less as game. Yet Elliott Barker, ranger under Leopold on the Carson, later professional mountainlion hunter, and then state game warden of New Mexico, has suggested in retrospect that Leopold may never have been fundamentally committed to the complete extermination of predators, even during his years on the Carson, where they were a real menace.[27]

Of the three major public campaigns for game protection in which Leopold assumed leadership during 1915–1920, the one in which he became perhaps most deeply involved was the campaign for a system of federal game refuges on the national forests. In this he had undoubtedly been inspired by the zeal of William T. Hornaday, although Leopold, unlike Hornaday, regarded the refuge as a device to produce game for hunting, not primarily to protect it from hunting. The bill that both Hornaday and Leopold promoted in 1915 was actually a plan for a large number of relatively small refuges, ten miles square or so, scattered throughout the forests and interspersed with hunting grounds. It was a far cry from Hornaday's original idea to turn the entire acreage of the national forests into inviolate sanctuaries like the parks.[28]

26. "The Game Situation in the Southwest," p. 5.
27. Barker's observation was made in a conversation 7 April 1966.
28. The Boone and Crockett refuge plan is discussed in James B. Trefethen, *Crusade for Wildlife* (Harrisburg, Pa., 1961),

Although legislation to establish a system of refuges had failed repeatedly, the Congress had established a number of large sanctuaries to preserve remnants of vanishing species. One such area was the million-acre Grand Canyon National Game Preserve, established in 1906 to prevent the extermination of a herd of about four thousand Rocky Mountain mule deer on the Kaibab Plateau north of the Grand Canyon. Other large preserves were established by the states on the national forests, especially on the elk ranges of Montana, Wyoming, and Idaho. But by 1915 there were indications that large, inviolate sanctuaries could create problems as well as solve them. Elk were already a cause célèbre in Yellowstone National Park and in refuges on the surrounding national forests. A growing herd of some fifty thousand was in danger of starvation because most of their formerly extensive natural winter range had been appropriated by ranchers and fenced. The elk in turn were seriously depleting winter forage required by domestic livestock and also by bison, bighorn sheep, and deer, making for a tangle of conflicting interests among ranchers, hunters, and wildlife preservationists. Thus Forest Service officials in Washington were beginning to question the concept of large sanctuaries, whether established by Congress or by the states, and to advocate greater administrative discretion for the Service in establishing refuges and determining the amount of game that should be harvested each year. Such discretion of course, would require not only congressional authorization but also a careful study of game and game lands by the Biological Survey and cooperative agreements with the states.[29]

---

pp. 63–72; Hornaday's efforts are chronicled in a number of his own books, especially *Thirty Years War for Wild Life: Gains and Losses in the Thankless Task* (Stamford, Conn., 1931), pp. 213–19.

29. See especially an article by Chief Forester Henry S. Graves, "The National Forests and Wild Life," *Recreation* (May 1915), 236–39. Although Graves' ultimate concern may have been more to protect the Forest Service's administrative discretion than to protect wildlife, his interest in moving forward in the wildlife field must have been an important factor in creating an atmosphere in which Leopold could function effectively.

Out in the Southwest, Aldo Leopold was not going to wait for congressional authorization or for studies by the Biological Survey. In addition to sparking the public campaign for the refuge bill, he got the cooperation of local forest rangers, sportsmen, and stockmen in making a preliminary survey of game populations and game ranges and began working up a proposed system of refuges, ready to be implemented as soon as the federal enabling legislation passed. The bill failed, and during the war a number of states again began establishing state refuges on the national forests. Leopold was particularly dismayed with the situation in Arizona, where entire isolated ranges of mountains, including his old stamping grounds in the Blue Range, were declared refuges, thus precluding any hunting. Such distortions of the refuge idea, he felt, could discredit the whole effort, especially since they often had a political flavor.[30]

When refuge bills were again introduced in Congress during 1919–1920, Leopold wrote a number of articles explaining and pleading for his kind of refuge—the refuge as a device for producing an outflow of game for hunting. A refuge was not a refuge, he asserted, unless it was surrounded by hunting grounds, freed from vermin, and provided with necessities such as food plants, water, fences, salt to make the breeding stock as productive as possible; it was not a refuge unless it was enforced, and unless the boundaries were both well marked and permanent. It is clear that as far as national forest refuges were concerned, Leopold was thinking mostly about production of deer and turkeys and saw refuges as the only alternative to closed seasons. He acknowledged one function of refuges —*sanctuaries* he termed them—for protecting endangered species such as mountain sheep and ptarmigan, but he did not dwell on that function. Moreover, he carefully exempted elk from consideration, lest their tendency to con-

30. Documents pertaining to Leopold's proposed system of refuges have not been found, but the progress of his campaign can be followed in daily bulletins and reports of District 3 and in the *Pine Cone* and other newspapers. For his attitude toward the Arizona refuges see "Wanted—National Forest Game Refuges" and other articles. (He did not specifically mention the Blue Range Refuge.)

flict with livestock interests be used as an argument against his plan. Elk should in general be confined to national parks, he thought.[31]

With respect to deer, Leopold explained, refuges had a very special function: They would solve the perennial problem of buck shortage. Most states allowed the killing of antlered deer only, a restriction of which he thoroughly approved. But where hunting pressure was severe, a shortage of bucks developed, and the available does tended to be bred, if at all, by immature males; this situation could eventually result in actual physical deterioration of the herd. Within the refuges, however, there would be a constant excess of prime bucks, deer being polygamous. The strongest bucks would appropriate the does on the protected refuge range and crowd the excess bucks, about three-fourths of the total, onto the surrounding range, where they would not only redress the buck shortage for hunting purposes but also breed heretofore unbred does. Not only would this mean greater herd productivity and vigor, but it would also effectively quell the argument for an open season on does that arose whenever hunters began seeing all does and no shootable antlers. So went Leopold's theory as of about 1920.

Policymakers in the Washington offices of both the Forest Service and the Biological Survey apparently felt that Leopold was overstating the case for refuges and might be giving the impression in his articles that refuges were more important than other aspects of game policy, such as plans for a limited license system. Yet at the same time as an editor for the Biological Survey characterized one of Leopold's articles on refuges as "a careless, sloppy piece

31. See especially "Wanted—National Forest Game Refuges;" "What is a Refuge?" *All Outdoors,* 8:2 (Nov. 1920), 46–47; "A Complaint," *The Game Breeder* (c. 1920), 288–89. As for Leopold's attitude toward restocking with elk, records in NA–RG95, Division of Information and Education, indicate that most of Leopold's more extreme statements against elk were toned down or edited out of his articles in the official Forest Service– Biological Survey review process at the office in Washington, D. C. In a letter to John Burnham of the American Game Protective Association, 20 Dec. 1919, Leopold stated flatly, "We want no more elk in the Southwest."

of work" lacking in restraint, publication of which would serve only to stir up muddy waters, the Chief of the Survey, E. W. Nelson, recognized Leopold's enthusiasm for the refuge idea by urging him to accept a position as head of the Survey's refuge program. Many of the refuges administered by the Biological Survey in those days, however, were special preserves for rare and endangered species or for the troublesome Yellowstone elk, not the sort of refuges that most interested Leopold.[32]

The bill providing general authorization for a system of federal refuges on the national forests never passed the Congress, but Leopold's efforts were not entirely in vain. He had promoted the idea so carefully and had drawn so much support at the local as well as the state level that when the new State Game and Fish Commission of New Mexico was created in 1921 it immediately began establishing a statewide system of refuges, generally following the carefully devised plan he had worked out years before. It was a system of which he was terribly proud.

In each of his three major campaigns for game protection in New Mexico—for law enforcement, for predator control, and for refuges—Leopold argued his case in the public arena and was eminently successful in winning support. By 1921 New Mexico, not yet a decade removed from territorial status, was recognized nationally by conservation leaders and government officials as one of the most progressive states in conservation affairs, and Aldo Leopold was regarded as the individual singly most responsible.

Game Management: The Science

In the years of the game protection campaigns, about 1915–1920, Leopold had regarded himself as a propagandist, an enthusiast in the "cause" of game conservation. But even during those years, he was reaching beyond

32. See correspondence in NA–RG95, Division of Information and Education, especially E. W. Nelson to the Forester, 12 Jan. 1920; H. A. Smith to District Forester, 24 Jan. 1920; LCE to Smith, 9 Jan. 1920. Nelson to AL, 27 Dec. 1919, offering him the USBS position, is in LP 11M1, but there is no reply. (Leopold's "Wanted—National Forest Game Refuges" was submitted for Washington office review on 13 Dec. 1919.)

propaganda for the beginnings of a science of game management. The new science, he believed, should be developed by foresters to make game a major forest product, just as foresters had developed the sciences of timber and range management. Forestry itself had started out as a cause, he observed in an article on "Forestry and Game Conservation" in 1918, but once the propagandists had won from the American people an affirmative answer to the question of *whether* the forests ought to be conserved, the science of forestry came to the fore prepared to deal with the question of *how* the forests should be conserved. The American people had already indicated their desire for game conservation, it seemed to Leopold, but a science of game management was not yet in sight; and foresters, who had been quick to see their responsibility and their opportunity in the related field of range management, were anything but aggressive when it came to a program for game.[33]

The forester as technician was concerned chiefly with timber, but the forester in his capacity as land manager, Leopold believed, was responsible for putting the land to its highest use. That involved provision for recreational hunting as well as for the harvesting of timber and the grazing of livestock. To Leopold, the doctrine of highest use—as enunciated by Gifford Pinchot, "that all land is to be devoted to its most productive use for the permanent good of the whole people"—apparently implied that the greatest possible variety of land uses should be maintained. It implied a recognition that conflicts among certain uses were inevitable and a responsibility to adjust these conflicts in the context of over-all policy so that all uses would be provided for, not necessarily on the same lands but on the most suitable lands and roughly in proportion to their relative social importance and the number of users. Foresters, Leopold explained, had been dragging their heels on game matters for three reasons, none of which satisfied him: first, the matter of dual authority, with the federal government owning the land and the states having juris-

33. "Forestry and Game Conservation," *Journal of Forestry*, 16 (April 1918), 404–11. The quotations on forestry and game management on the next few pages are drawn from this article.

diction over the game; second, the lack of a strong local demand for game conservation—until Leopold, following the lead of earlier foresters with respect to grazing administration, demonstrated in New Mexico that it was "quite possible to deliberately go out and create" such a demand; and third, "the vague fear that a real crop of game might interfere with both grazing and silviculture, as if grazing and silviculture might not also interfere with each other!" To this quip he added, "The principle of 'highest use' has evidently been more talked about than understood."[34]

Leopold not only believed that foresters had a special responsibility to handle the game problem in order to protect the recreational value of the forests, but he also thought that the new science of game management could borrow a great deal from the science of forestry. He suggested an analogy between management techniques that could be applied to big game such as deer and sustained-yield forestry. Game census could be compared to timber reconnaissance, protection against predatory animals and illegal kill to fire protection and timber trespass cases, breeding stock to growing stock, hunting demand to timber market, limitation of kill to limitation of cut, game laws and license fees to sale contracts and stumpage rates, natural increase and artificial restocking to natural reproduction and planting, and so on.

34. Pinchot's statement appears in a letter written to himself and signed by Secretary of Agriculture Wilson, 1 Feb. 1905, which refers specifically to use of wood, water, and forage. The principle of highest use has been transmuted over the years into the current Forest Service principle of 'multiple use', given statutory basis in the Multiple Use-Sustained Yield Act of 1960, which specifically cites outdoor recreation, range, timber, watersheds, and wildlife and fish. The report of the Public Land Law Review Commission, *One Third of the Nation's Land* (Washington, D.C., 1970), states that "management of public lands should recognize the highest and best use of particular areas of land as dominant over other uses." In view of other recommendations that timber production be "dominant" on certain lands, some have seen in this resort to 'highest use' terminology an attempt to unshackle the Forest Service from the equal attention to wildlife and recreation statutorily imposed by the Multiple Use Act (but originally argued for by Leopold, it may be noted, under the principle of highest use).

There were two points at which Leopold felt compelled to make a distinction between game management and forestry. One was the matter of selection of species, a point that in his estimation could not be overemphasized:

> Forestry might prescribe for a certain area either a mixed stand or a pure one. But game management should always prescribe a mixed stand—that is, the perpetuation of every indigenous species. Variety in game is quite as valuable as quantity. In the Southwest, for instance, we want not only to raise a maximum number of mule deer and turkey, but we must also at least perpetuate the Mexican mountain-sheep, bighorn, antelope, white-tail deer, Sonora deer, elk, and javelina. The attractiveness, and hence the value of our Forests as hunting grounds, is easily doubled by retaining our extraordinary variety of native big game. This variety also adds enormously to their attractiveness for the summer camper, the cottager, and the fisherman. The perpetuation of interesting species is good business, and their extermination, in the mind of the conservationists, would be a sin against future generations.

There are several things to note about this early argument for preservation of variety. First, it does not extend to predatory species. Second, Leopold's concern was not to maintain balanced populations of all species but to produce maximum deer and turkey (the commercial species) and simply to avoid extinction of the rest. Extinction, it seemed to him, was an imminent threat to game species but not to tree species ("Black walnut or yellow poplar may have become commercially defunct in our hardwood forests, but they are not extinct and never will be"). His grounds for preserving diversity at this time were economic, esthetic, and moral, rather than ecological.

The other point at which the analogy between forestry and game conservation diverged was on the matter of inaccessibility, but here the difference was all in favor of game. American foresters might have preached sustained-yield forestry, but they had as yet been unable actually to practice it, he acknowledged, *"because of lack of a demand* for inferior grades and remote stumpage. Because of our old bugbear—inaccessibility." Did game management la-

bor under the same handicap of inaccessibility? "Emphatically it does not," he asserted, *"There is a demand for every head of killable big game in the United States,* wherever it may be." So far as the economics of supply and demand were concerned, we were more ready to practice game management than forestry.

The biggest problem in applying sustained-yield principles to game was the lack of any volume limitation on annual kill, except for the one-buck law, which made no stipulation on the number of hunters. Leopold accordingly put forth a plan for a system of federal hunting permits to be issued annually by management areas, such as ranger districts. If the federal permit were required in addition to a state license and its use were subject to state seasons and other hunting laws, he did not see any grounds for conflict over jurisdiction. Leopold's plan, it may be noted, would work admirably when the aim was to restrict the kill in the face of excessively liberal state laws, but it would not provide a means of augmenting the kill when state laws were too limiting. That he did not even consider the possibility of having to augment the kill indicates that he was still somewhat bound by the restrictive mode of thinking in the game conservation sequence of ideas, even though he was fully committed to the productive idea and to the application of scientifically determined controls.

In a subsequent article, "Determining the Kill Factor for Blacktail Deer in the Southwest," Leopold described the 'kill factor' formula he and his colleagues in District 3 hoped to use to determine scientifically the optimum annual kill from a particular area of deer range. The formula was analogous to the 'steer factor' used in range cattle management, which was in turn based on the concepts of annual increment and yield tables in sustained-yield forestry. The steer factor, which had been derived empirically over the years and had come to be accepted by stockmen, forest officers, and bankers alike, indicated that under normal conditions the number of mature cattle on the range (the unknown, which the Forest Service needed to estimate in order to determine proper stocking and issue grazing permits) was equal to about six times the number of steers sold annually. The kill factor in deer management, similarly, would be the ratio between the number of mature deer in an area and the number of bucks that could be killed there annually without decreasing the breeding stock; as such it was roughly a measure of herd productivity on normal range, with normal factors bearing on the productive capacity of the herd, such as losses due to predatory animals, illegal kill, starvation, and disease, automatically taken into account. It could be used to determine the allowable kill when a census had been taken, or the census when the hunter kill was known.[35]

As early as 1914 Leopold had had the ranger in the Magdalena Mountains in the Datil National Forest of west-central New Mexico keep accurate records of deer census, annual kill, and predatory animal losses, thinking that the Magdalenas would make a good sample plot on which to determine productivity. Although the records were apparently lost during World War I, Leopold intended to resume the collection of data in the Magdalenas and in other sample areas of varying conditions in the Southwest in order to derive a series of kill factors. What became of the project is not known, but he did not lose interest in deer and kill factors.[36]

35. "Determining the Kill Factor for Blacktail Deer in the Southwest," *Journal of Forestry*, 18:2 (Feb. 1920), 131–34. From the rough census and kill estimates submitted by forest rangers since about 1915, Leopold calculated a tentative average kill factor of about 1:10 or 1:15, indicating a considerably lower productive capacity for deer than for cattle. This seemed reasonable to him, in view of the longer time it took to produce a killable buck than a steer, the less complete control of mountain lions (which preyed mostly on fawns) than of wolves (which preferred calves), and the "lowered specific resistance" of deer herds in their decimated state. On the basis of accurate kill records and his own best estimate of deer census on the Gila National Forest in the mid-1920s, he derived a conjectural kill factor of only 1:35, which seemed too low. Because of the uncertainty of the census estimates, he decided not to use the census-based kill factor, but to rely instead on a comparison between area and kill. See "Southwestern Game Fields," Ch. IV, p. 8; and this text, Ch. 3.

36. Leopold recounted his early interest in the Magdalenas in a letter to F. L. Kirby, Forest Supervisor of the Datil, 27 Dec. 1926, LP 4B1.

Southwestern Deer and the Concept of Productivity

The new concepts were utilized in "Southwestern Game Fields" on which Leopold began working some time in the early 1920s with J. Stokley Ligon of the U.S. Biological Survey and R. Fred Pettit, a dentist from Albuquerque. All three set about making observations and gathering data from hunters, forest rangers, naturalists, and other outdoorsmen; but it was apparently Leopold who supplied the conceptual framework, based on the analogies with forestry and range cattle management that he had begun to work out in his early published articles. He also did most of the writing, at least of those portions of the manuscript known to be extant. By spring of 1927, three years after he had left the Southwest to become associate director of the Forest Products Laboratory in Madison, Wisconsin, he had preliminary drafts of five chapters of "Southwestern Game Fields" ready to send around to his colleagues for criticism.

It was an ambitious project. Not only did it aim to present a new approach to game conservation based on an understanding of population mechanisms and techniques for manipulating them, but it also aspired to be a regional manual, describing the game of the Southwest in terms of local environmental conditions and prospects for management, and a species monograph, demonstrating the new approach as it might be applied to management of deer in the Southwest. It was too ambitious for a single volume. Later that year, Ligon incorporated much of their data on life histories and distribution of various game animals in his *Wildlife of New Mexico*, published by the New Mexico Department of Game and Fish, where he was now employed as a game specialist. Leopold then outlined a new book, to be focused more narrowly on "Deer Management in the Southwest." By June 1929 he had redrafted a few chapters, and Ligon had prepared several short sections assigned to him.[37] Thereafter the project was aban-

37. Leopold's drafts are in LP 6B10. The few sections prepared by Ligon (included in LP 6B10 and also with the Ligon Papers in the Conservation Library Center at the Denver Pub-

doned, for reasons that can only be surmised: communication problems among the authors, lack of data to substantiate the concepts they were formulating, Leopold's intention to write a more general text on principles of game management. Probably most important were events on the deer ranges, which called into question many of the assumptions on which Leopold's approach was based at a time when he was not on the ground to work them out.

The key concept in Leopold's manuscripts about the Southwest was productivity, which he defined as the rate at which mature breeding stock produces other mature stock, or mature removable crop. He pictured deer population as an exponentially rising curve of unimpeded natural increase that was pulled down to horizontal equilibrium, in a stable herd, by a number of 'factors of productivity' (Figure 1). Some of the factors, termed *decimating factors,* killed directly—hunting, predators, starvation, disease, and accidents—while others, termed *welfare factors* —poor food, poor water, poor coverts, and special factors— tended to decrease the breeding rate or weaken an animal's physical resistance to the decimating factors. Productivity, considered as a measure of huntable surplus, could be increased by a simple transfer of mortality from predators

---

lic Library) are the only extant evidence of contributions to the writing by either Ligon or Pettit. Leopold's outlines for "Deer Management in the Southwest," however, include references to various revisions planned for "old chapters" I–XII. Because "Southwestern Game Fields" includes parts of only five chapters, there may well have been an intermediate draft, or substantial additions to the first draft, which could have been written by any of the three authors. Most of Leopold's correspondence pertaining to the manuscripts is in LP 4B1-8, but there are frequent references to materials that Leopold was sending to Pettit or Ligon "for the files." It is not clear whether these materials are now in LP 4B1–8, or not. There are a number of "checklists of life histories" of various species, apparently prepared by Leopold from notes submitted by Ligon—some of which may have been incorporated in *Wildlife of New Mexico.* However, there is almost nothing in Leopold's files that could possibly be attributed to Pettit. (It is possible that heirs of Pettit and Ligon who have not been located may have some relevant materials.)

and other decimating factors to the hunting factor, or by controlling the welfare factors and thereby increasing the breeding rate. Neither would necessarily increase the population level if the increment were removed by hunting.

Implicit in "Southwestern Game Fields" and explicit in "Deer Management in the Southwest" was a distinction between the nine fundamental factors of productivity and

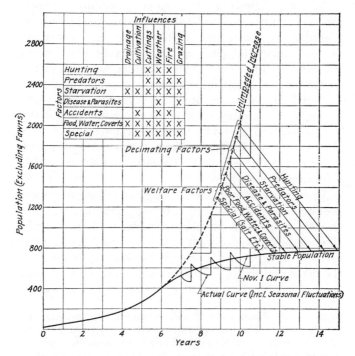

1. Factors of Productivity in Relation to Population (Mule Deer). (From "Deer Management in the Southwest.")

infinite variety of 'environmental influences', such as drainage, cultivation, cuttings, weather, fire, and grazing, which conditioned productivity indirectly by influencing one or more of the factors. Because these environmental influences could usually be reduced to factors, as common denominators, Leopold felt justified in focusing exclusively on the factors in his discussion of environmental control.

73

He went so far as to write, "The factors collectively constitute the environment."[38] As we have already seen, he had a difficult time in his chapter "The Virgin Southwest and What the White Man Has Done to It" evaluating the over-all effect of environmental change on game; this is undoubtedly one of the reasons he chose to focus on factors rather than on the more complex interactions of influences.

As in his analysis of equilibrium or stability in watersheds, Leopold recognized that population curves, even on virgin terrain, were never static but tended to oscillate about a horizontal median. This dynamic equilibrium between the breeding potential of the species and the environmental resistance (the factors of productivity, collectively) had, however, been thoroughly upset by civilization. Game management, as he conceived it, was the "deliberate and purposeful manipulation of the factors determining productivity"; its purpose was "to substitute a new and objective equilibrium for the natural one which civilization has destroyed."[39]

Again as with watersheds, Leopold recognized many possible levels of equilibrium in game populations, from the relatively sparse populations under wilderness conditions to the more exuberant levels resulting from an interspersion of agricultural and livestock developments, to the dense populations maintained on game farms and in some of the European countries by more or less completely

38. "Deer Management in the Southwest," c. 1929, Ch. III, p. 4, LP 6B10. The conception of productivity presented here is drawn from "Deer Management in the Southwest," which is refined somewhat from that in "Southwestern Game Fields" but essentially the same as that presented by Leopold in *Game Management*. Leopold's conception of factors was apparently influenced by Royal N. Chapman, "The Quantitative Analysis of Environmental Factors," *Ecology*, 9:2 (April 1928), 111–22. In this article, however, Chapman did not distinguish between decimating and welfare factors or mention the concept of influences. For further discussion of factors and influences see pp. 118–19.

39. "Southwestern Game Fields," Ch. I, pp. 3, 24. The comparison of population equilibrium and watershed equilibrium was not made by Leopold at the time, but see pp. 178–79 of this text for an analogy he drew years later.

artificialized game management. It would, however, have been esthetically objectionable to him, and he felt to Americans as a people, to artificialize the production of game, as in Europe, by intensive control of virtually all the factors of productivity, especially when America was blessed with such an abundance of land. What was least artificial was to him most esthetically if not yet most ecologically desirable. Hence, he presented management as a matter of applying science in determining optimum levels of population for American conditions and as a matter of skill in selecting only the most crucial factors (the limiting factors) to manipulate.

The theory of productivity and its implications for management seemed so important to Leopold that even as he redrafted chapters for "Deer Management in the Southwest" he was already planning to devote a separate future volume (*Game Management*, 1933) to a further elaboration of the subject. In the meantime he would work out the theory and its implications for management of just one game animal, deer, in a single geographic region.

3

The Gila Experience

The Gila as Normal Range

Through his manuscripts about the Southwest, Leopold intended to demonstrate methods for determining and achieving "a new and objective equilibrium" in southwestern deer populations. He would deal with the question: What is proper stocking and productivity? From forestry he borrowed the concept of 'normal stocking', which involved comparison with measured standards that were representative of the best-known stocking of various species on different soil types. In deer management, similarly, he would select the most productive, stable herd of deer that was typical of a given environment, analyze it, and use it as a standard against which to measure other herds on similar ranges.

The best-stocked deer ranges Leopold knew were the first ones he had encountered when he arrived in the Southwest, the mesa-and-canyon terrain at the eastern end of the Mogollon Rim, on the Apache, Datil, and Gila national forests. He might have selected the Blue Range on the Apache as his example of a normal range, if the entire 600,000 acres had not been declared a refuge by the rather too-political, too-volatile game administration in the state of Arizona. He had early sought to gather data on deer in the Magdalena Mountains of the Datil Forest, as we have seen, but later apparently decided the area was too small and there was too much drift of deer to and from surrounding terrain to make it a reliable sample. Consequently he chose to use the 2,500-square-mile Gila National Forest in southwestern New Mexico, east of the Apache and south of the Datil, as his standard of normality (Figure 2).[1]

1. The Datil National Forest was discontinued as a separate unit in 1931 and divided among the Apache (north and west), Gila (south), and Cibola (north and east). The Magdalena Di-

Like the Apache and the Datil, the Gila was typical southwestern deer country, historically well stocked; it had been saved from depletion by the presence of Geronimo's Apaches until the mid-1880s, but merciless overgrazing, severe flooding and erosion, and market hunting to feed the early miners and railroad crews took their toll rapidly after that. One hunter, operating during the winter of 1888–1889 from the N-Bar Ranch near Negrito Mountain, delivered at Magdalena 387 carcasses ranging up to 237 pounds drawn and headless—a measure of early-day abundance of deer as well as the pressure that could be exerted on them by market hunters. Allowing time for packing sixty miles to market, Leopold figured the man must have killed an average of six deer a day; and that, he commented, is approaching "rocking-chairs behind every bush."[2]

Leopold became interested in the Gila in part through family and personal connections. The N-Bar Ranch had been acquired by his wife Estella's uncle, Solomon Luna, one of the most successful sheepmen and most influential political figures in the territory of New Mexico. He used it as a base for his sheep-grazing operations on the eastern slope of the Mogollon Mountains at the headwaters of the west and middle forks of the Gila River and out on the open, rolling Plains of San Augustin. In 1912 Estella Leopold's mother inherited the ranch and it was managed by her half-brother Eduardo Otero, who as president of the New Mexico Woolgrowers' Association was enormously helpful to Leopold in winning stockmen's support in the early years of the game protection campaign. The president of the New Mexico Cattle Growers' Association, Victor Culbertson, manager of the largest cattle outfit on the Gila, the GOS, was also a friend of Leopold's and helpful in the cause, as were two other prominent cattlemen on the forest, Hugh Hodge of the Diamond Bar and G. W. Evans. Fred Winn, who had worked with Leopold on the

---

vision (about 100,000 acres) went to the Cibola; the N-Bar Ranch, mentioned above, to the Gila.

2. "Southwestern Game Fields," typescript, c. 1927, Ch. III, p. 8, LP 6B10.

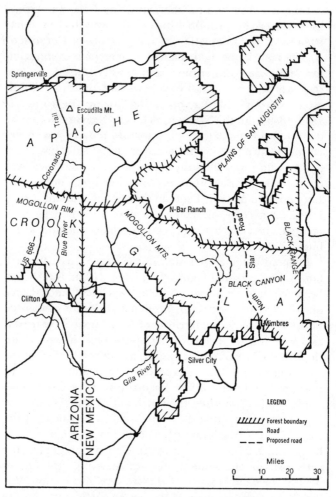

2. *Gila Headwaters Area, 1929.*

Apache, was appointed supervisor of the Gila Forest in 1916.

The Sportsmen's Association of Southwestern New Mexico, first such organization in the state, was formed in Silver City, headquarters of the Gila Forest, in 1914, perhaps even before Leopold himself took up the cause. Leopold began almost immediately to work with leaders of the association, with Winn, and with his stockman friends to mount a campaign for better sportsmanship and enforcement of game laws. Together they secured the services of J. Stokley Ligon and the best lion hunter of them all, Ben Lilly, in routing predators from the Gila ranges. They also determined the boundaries for a series of deer refuges on the forest. At the same time, Fred Winn made a determined effort to reduce stocking of cattle and sheep on the ranges, although hindered somewhat by wartime demands for meat. By the early 1920s, deer were rapidly increasing on the forest; and some of the largest grazing allotments, such as those assigned to Ed Otero and to the GOS and Diamond Bar cattle outfits, were coming to be regarded as models of conservative range management.

Leopold was interested in the Gila not only as a result of early familiarity, personal friendships, and local success in organizing sportsmen and securing increased deer populations but also because he recognized its potential for a new type of forest land use. By 1921, if not earlier,[3] he had begun arguing for designation of the Gila as a national hunting ground, or wilderness area, by which he meant "a continuous stretch of country preserved in its natural state, open to lawful hunting and fishing, big enough to absorb a two weeks' pack trip, and kept devoid of roads, artificial trails, cottages, or other works of man." With the Coronado Trail about to bisect the Apache Forest, the Gila would be one of the last remaining roadless areas of more than a half million acres in the Southwest. Leopold argued for its designation as wilderness on the principle of 'highest use', the same principle on which he had argued for game management on the national forests. In this case, however, he was arguing not for preservation of variety in game but for preservation of a distinctive kind of recreational hunting experience, the pack trip, unique because opportunities for it were disappearing almost everywhere else and vital as a sample of an experience

3. Mary Orr Russell of Reserve, New Mexico, recalls that Leopold was interested in the wilder areas of the Gila when she first knew him during his years as a forest assistant on the Apache. Conversation with author, 2 June 1974.

important in the development of the national character. He was arguing for preservation of diversity in recreational and cultural experience.[4]

Leopold made an official inspection of the forest in 1922, in connection with which he worked up a proposal for administration of the Gila as wilderness and drew a tentative boundary line encompassing over a million acres. A year later he and his wife packed into the Black Canyon–Diamond Creeks area in the central Gila for a memorable two-week deer hunt. They jumped deer every day, both does and bucks; his wife recalls having counted thirty in a single group. Leopold brought down an eight-point buck on that trip, one of only two deer for which there is evidence of his ever having killed (the other was taken for camp meat during the 1909 Blue Range reconnaissance).[5]

Quite possibly at Leopold's instigation, the Sportsmen's Association of Southwestern New Mexico, in cooperation with the Forest Service, the Biological Survey, and the state, inaugurated in 1923 an annual check of all hunters entering or leaving the Gila Forest, in order to get an accurate count of deer killed and an estimate of proportions of bucks and does seen on the forest. In 1927 Leopold believed that these statistics constituted the best continuous record of deer populations anywhere in the Southwest and

4. "The Wilderness and Its Place in Forest Recreational Policy," *Journal of Forestry*, 19:7 (Nov. 1921), 718–21. Leopold noted that Pinchot enunciated his doctrine of 'highest use' in order to emphasize that the national forests would be opened up and developed as producing lands, at a time when many special interests opposed the creation of the forests for fear they would be locked up as wilderness. The process of development "must of course, continue indefinitely," Leopold observed, but added that it had already gone far enough to raise the question of "whether the principle of highest use does not itself demand that representative portions of some forests be preserved as wilderness."

5. Leopold's "Report on Proposed Wilderness Area," typescript, 2 Oct. 1922, 3 pp., is on file at forest headquarters in Silver City. The full inspection report, however, has apparently been lost. The 1923 hunt is recounted in Leopold's "New Mexico Journal," LP 7B2. The reference to the 1909 deer, "a fat blacktail buck," is in AL to Cicero (his sister Marie), 4 Oct. 1909, courtesy of Marie Leopold Lord.

perhaps in the nation—a commentary on the status of game statistics in those days. It was the availability of these statistics, coupled with the quality of the deer range, the apparent effectiveness of law enforcement, predator control, and refuges in restoring the depleted herd, and Leopold's own personal familiarity with the forest and with local citizens that led him to focus on the Gila in "Southwestern Game Fields."

Designation of the central portion of the Gila as wilderness in 1924 and Leopold's selection of it as 'normal range' for his deer study would lend a certain ironic authenticity to his argument years later for wilderness as a base datum for science.

Black Canyon and the Kaibab

In August 1927, three months after Leopold sent the first draft of "Southwestern Game Fields" around to his colleagues for criticism, his coauthor J. Stokley Ligon was dispatched by the New Mexico Department of Game and Fish to investigate reports of overabundance of deer and damage to forage in the Black Canyon area of the upper Gila drainage. Ligon reported severe depletion of forage over at least 100 square miles of range and numerous deer over several hundred square miles. Even if all cattle were removed from the conservatively stocked Diamond Bar grazing allotment, he indicated, the supply of forage would be inadequate for so many deer. From an actual sample count, he estimated there were 38 deer per square mile over much of the upper Gila drainage, up to 43 in some sections, and he noted that concentrations would be even more dense in the winter when the big bucks came down from the high country. There was a noticeable shortage of fawns, however, and most of the does and immature animals appeared in poor physical condition owing, Ligon surmised, to poor forage conditions. Deer were depleting their supply of preferred forage species like mountain mahogany and live oak and were turning to alligator juniper, which fortunately had become markedly more abundant in the area in recent years. But even the junipers, which cattle scarcely touched, were browsed as high as deer could reach, and there was evidence the deer had fed extensively on the branches of pinyon pine, which they would eat only in extreme necessity. Conditions were worst in the 16,000-acre Black Canyon Game Refuge, which served as a concentrating ground for deer in winter. Although he realized it would be entirely contrary to game refuge policy, Ligon recommended that the refuge be opened to hunters during the coming season in order not only to reduce the surplus but also to scatter the animals, rather than force them into further congestion by hunting pressure outside. The game department should take steps, he advised, to direct as many as five hundred hunters into the upper Gila drainage that fall, and the game commission should be vested by the legislature with authority to deal fully with such emergencies.[6]

Developments involving the Black Canyon refuge were of consequence for Leopold, who had drawn the boundaries for it back in February 1916 and hunted the theoretical "outflow" of deer from it in 1923. He was proud of the system of refuges he had been instrumental in establishing in New Mexico and justifiably anxious that it not be dismantled. When earliest reports of excess deer had reached him in February 1926, he had written immediately to the new supervisor of the Gila Forest, J. A. Scott, volunteering his opinion that the relative inaccessibility of the Black Canyon country to motorized hunters might be a factor contributing to the increase. He suggested that if the load on the range had to be lightened, the refuge should be decreased in size rather than abolished.[7] Even this admission, that an overstocked refuge in an inaccessible area might eventually have to be smaller, marked a considerable shift

6. J. Stokley Ligon, "Preliminary Report—Gila Deer Situation," typescript, 11 Aug. 1927, 9 pp., LP 3B8.

7. AL to James A. Scott, 26 Feb. 1926, LP 4B1. J. Stokley Ligon had begun rethinking the refuge question in his capacity as game specialist for the state game department, and wrote Leopold on 26 March 1926 requesting his advice concerning the large refuges in inaccessible country. Leopold replied 6 April that he did not think refuges had been too large in New Mexico in the past. Theoretically they might have been smaller and more numerous, he said, but for economy of patrol the larger size was justified.

in his thinking on the matter of inaccessibility since the days when he had insisted, "There is a demand for every head of killable big game in the United States, wherever it may be."

Although we do not know Leopold's specific reaction to Ligon's report, which he was sent immediately, we can imagine he was astounded at the estimates of deer density, upwards of 38 per square mile. In his chapter on "Normal Deer Stocking and Productivity" in "Southwestern Game Fields" he had assumed, for want of a reliable census, that the Gila averaged five deer per square mile and ventured the guess that the best large western ranges, in California, did not run more than ten. Leopold did not trust most census estimates for game, however, because they were too variable as to the individual making the estimate and the amount of "blanks" in the area he was estimating for. Without an accurate census, he could not establish a kill factor. To arrive at a reliable standard of normal stocking and productivity in his manuscript, therefore, he had used the concept of 'area:kill ratio', or bucks killed by hunters per square mile of deer range per year. The total kill was known for the Gila for the years 1923–1926 (about 362 bucks a year) and averaged to about 0.15 bucks per square mile per year, which Leopold therefore took as normal for areas of similar size and environmental conditions in the Southwest. (The Jemez and Pecos areas of the Santa Fe National Forest, by contrast, had an area:kill ratio of 0.015, only one-tenth as productive as the Gila, although these areas reputedly had once been good deer country.)

The joker hiding in Leopold's calculations was his assumption of stability. Indeed the whole concept of normal stocking and productivity was based on an assumption of sustained yield, or a stable population, and the Gila herd was anything but stable. Leopold, back in Madison, received the annual kill statistics for the Gila and worked out the area:kill ratio, but these figures alone would have provided no hint that the deer population was mushrooming. The kill ratio was based on hunter kill, and in the inaccessible Gila the number of hunters, and therefore the total kill remained relatively constant, even though the supply of deer increased phenomenally. Leopold seems not to have appreciated this factor at the time he was writing "Southwestern Game Fields," perhaps because he was still thinking of hunter demand as relatively elastic, inaccessibility notwithstanding. Ligon's estimate, made by a thoroughly reliable game manager on the basis of a sample count, was thus the first real indication Leopold had of the astounding increase in density of deer on the Gila.

The one obvious instance of overpopulation in the Southwest prior to reports from Black Canyon was on the Kaibab National Forest in Arizona, a case that has since become the most celebrated example of irruption and collapse of a deer herd in wildlife literature. The population on the Kaibab had increased from about 4,000 in 1906, when the Grand Canyon National Game Preserve was established, to 30,000 in 1924, as estimated by the forest supervisor. (Years later, students of the Kaibab case would estimate the 1924 peak at 100,000.) Forest officers had reported as early as 1918 that the increasing herd was beginning to influence the supply of forage, but six successively more urgent warnings up to 1924 failed to provoke a program for significant reduction.[8]

The Kaibab deer were already recognized as a national asset, and the Forest Service was fearful the public would react violently toward any attempt to reduce the population. Millions of Americans had read a series of articles by Emerson Hough in the *Saturday Evening Post* urging elimination of sheep and cattle on the forest and preservation of the deer herd. It was "the greatest stag herd ever known in the United States," said Hough, arguing in 1922 for dedication of the Kaibab Plateau, "our last and noblest wilderness," as the President's Forest, to be set apart unchanged forever.[9] Meanwhile, most of the foods favored by deer like cliffrose, snowberry, and ceanothus were depleted, aspen, oak, and juniper trees were highlined, and

8. For a general history of the Kaibab herd see John P. Russo, *The Kaibab North Deer Herd—Its History, Problems and Management*, State of Arizona Game and Fish Department, Wildlife Bulletin No. 7 (July 1964).

9. Emerson Hough, "The President's Forest," *Saturday Evening Post*, 194:29 (14 Jan. 1922), 7, 72.

even white fir, usually eaten by deer only in times of stress, was heavily utilized. The herd as well as the range was in critical condition, and thousands of deer, including most of the new fawns, died of starvation during the severe winters of the mid-1920s.

In an effort to get broader sanction for action, the Secretary of Agriculture in 1924 appointed a Kaibab Deer Investigating Committee composed of national leaders of livestock, hunting, and preservationist organizations. The committee recommended immediate cessation of commercial livestock grazing on the forest and reduction of the deer herd by at least 50 per cent, but it weaseled on the means of herd reduction, suggesting trapping and shipping of deer to restock depleted ranges and only timidly broaching the possibility of some "shooting." After several miserable failures at transplanting deer, including an abortive attempt led by Zane Grey to drive deer down the Grand Canyon and across the Colorado River, the Forest Service that fall began issuing permits, three to a person, for the killing of deer on the Kaibab. The long-time governor of Arizona, George W. Hunt, a liberal activist fond of controversy and public exposure, promptly had three of the hunters arrested, thus making protection of the Kaibab deer a political issue and initiating what was to become a classic confrontation in the courts on the question of state vs. federal jurisdiction over wildlife on federal lands. The federal district court in Los Angeles in September 1926 sustained the right of the federal government to protect its property from damage by killing or removing as many deer as necessary, without regard to the game laws of Arizona, and the Santa Fe *New Mexican* gave the story a banner headline, "Government Can Kill Deer on Kaibab." Arizona appealed to the U.S. Supreme Court, but the case was still undecided and the Kaibab Plateau still woefully overpopulated at the time Leopold wrote "Southwestern Game Fields" and Ligon his report on Black Canyon.[10]

10. The Kaibab litigation and other aspects of the jurisdiction issue are ably treated in Ralph R. Hill, "Federal-State Jurisdiction Over Wildlife," typescript, Oct. 1968, 25 pp. (copy courtesy of author). See also James C. Foster, "The Deer of

Leopold was in Madison and no longer closely involved with the Kaibab problem, if indeed he ever had been. The Kaibab Forest was not administered by District 3 at the time, and Leopold apparently never visited the north rim of Grand Canyon or the Kaibab Plateau until 1941. As early as 1923, however, he had been assigned to work on the Arizona legislative end of the problem. Leopold was a great believer in encouraging the states to accept responsibility for handling their game, on the forests as well as outside them, but there is no evidence that he had any success with Arizona.[11]

Kaibab: Federal-State Conflict in Arizona," *Arizona and the West*, 12 (Autumn 1970), 255–68.

11. There is scant evidence of Leopold's involvement with the Kaibab problem in the early 1920s. A letter from E. A. Goldman to Leopold of 4 Jan. 1923 indicates that Leopold had earlier prepared a memorandum commenting on certain features of a report by Goldman on the Kaibab situation, but the memo has not been found. Leopold's legislative assignment is listed in the District 3 Program of Work for 1923 (National Archives), but there is no information as to what he may have done.

When the Forest Service and the state of Arizona reached a preliminary agreement in 1923 for federal operation of a hunting permit system on the Kaibab, Leopold expressed his disappointment in a letter to C. E. Rachford of the Washington office: "I am forced to consider how much better a strategic position we would be in if we had broached State control and the State had insisted on Federal control. Such a course would have taken the wind out of the sails of the States rights politicians. . . . It also cuts off the best opportunity to date for demonstrating the truth that in the long run the States will retain only such authority as they earn by their competency in game management." (AL to Rachford, 9 Nov. 1923, NA–RG95, Wildlife Management Corres., Box 16).

He also explained his position on encouraging state administration of game in correspondence during Dec. 1923–Jan. 1924 with George Bird Grinnell, president of the Boone and Crockett Club, who favored unconditional administration by the Forest Service of wildlife on the national forests. In Leopold's view, such a policy would tend to weaken state game departments; hence, "the thing to do is insist that the states handle their game, and to *take over the Forest game whenever or wherever they fail to do so*." (AL to Grinnell, 11 Dec. 1923, LP 2B2).

The Gila Experience

Although he was fifteen hundred miles removed from the scene of the action, Leopold recognized its relevance to his book on deer management, and he invited S. B. Locke, a forest examiner who had been acquainted with the Kaibab problem for years, to contribute a chapter on it. Locke may never have written the chapter, but Leopold referred to data collected on the Kaibab a number of times in his own chapters, usually to illustrate characteristics of overstocking. His treatment of the situation on the Kaibab, however, was entirely matter of fact, with no hint of alarm, no allusion to the perplexing political difficulties in trying to reduce excess deer, nor any dampening of his ardor for restoration of deer over the vast depleted ranges of the Southwest. The assumption was that management could deal as easily with problems of too many deer as of too few simply by increasing the allowable kill, once standards for normal stocking and productivity were determined. The causes and extent of the problem were not well understood at the time, so Leopold could believe that the Kaibab was a special case. A plateau bordered by the Grand Canyon and by desert, it was a biological island, virtually the entire extent of which had been closed to hunting upon establishment of the Grand Canyon Game Preserve in 1906. The area did not operate as a refuge, according to Leopold's canon, because it was too large and it was not surrounded by hunting grounds; the factor of hunter demand had had no chance whatever to function.

* * *

No doubt with the trials of the neighboring state in mind, the New Mexico game commission opened the Black Canyon refuge to hunting in fall 1927, though it had as yet no authority to increase the bag limit beyond the time-honored one buck. Aldo Leopold was planning a second packsaddle hunting expedition to the Gila that fall to test his new-found skill as an archer. It is hardly surprising that he avoided Black Canyon, heavily stocked with hunters, and concentrated on the middle fork of the Gila instead, going in via the N-Bar Ranch. He never saw such a hunting ground, he exclaimed in his journal—ten, twenty, and more deer, both blacktails and whitetails, virtually every day out, and turkeys up to forty in a flock. One big buck plunged away over a hogback with an arrow in him, but the rest successfully evaded Leopold's aim in sapling pine thickets that had grown prevalent in the area. Leopold observed that the year's pinyon crop was highgraded and the deer were feeding heavily on alligator juniper berries, but he gave no hint in his journal that he recognized impending trouble.[12]

On his return home he wrote an article, "Pineries and Deer on the Gila," in which he tried to account for the admirable recuperative capacity of the Gila deer, even in areas where they had been heavily hunted. Law enforcement, predator control, and refuges were in part responsible, of course, but another possible reason had impressed itself on him as he stalked deer on the middle fork—namely, all the sapling pine thickets, which made such fine cover for deer. As usual he had a theory. Severe overgrazing by cattle in the late nineteenth century had thinned out the grass on the area, and with the end of the period of heaviest grazing, pine (and in some places juniper) seeded in on the exposed soil on a large scale. The pine had survived in thickets, instead of being thinned out or destroyed, as a result of the virtual cessation of fires under Forest Service administration. It was a lucky accident that overgrazing, forestry, and fire protection should have combined to produce a habitat change favorable to deer. Even so, he warned, "Pines alone do not constitute game country." Although there had been a "radical improvement" in grass and herbage since he saw the area in 1922, the forage and the watersheds were still in precarious shape, and he expressed concern that with improved livestock markets the pressure to again overstock the ranges would be almost irresistible. "Sportsmen in New Mexico," he concluded, "will have to learn how to judge range problems, and help the more farsighted stockmen and the Forest Service resist the impending

12. This entry from Leopold's "Wisconsin Journal" has been published in *Round River: From the Journals of Aldo Leopold*, Luna B. Leopold, ed. (New York, 1953), pp. 96–105.

temptation to overstock. Game conservation without range conservation is an idle dream."[13]

"Pineries and Deer" represents Leopold's first published attempt to relate the processes of environmental change, about which he had written earlier in the decade, directly to the long-term fortunes of deer. It is curious that he should have rejoiced in the spread of pine and juniper thickets because in the long run these dense growths would inhibit reproduction of the sun-loving understory vegetation required by both deer and cattle as forage.[14] The article is, however, significant in its recognition of forage—a welfare factor—as the new limiting factor for deer on the Gila. No longer did Leopold emphasize the limiting role of decimating factors such as predators and illegal hunting. But the article is equally significant, as an indication of the state of Leopold's thinking, in its facile assumption that the future threat to deer forage would come from overstocking with cattle rather than with deer. Reports from Black Canyon and the Kaibab seemingly gave Leopold cause for concern about forage as the limiting factor for deer on the Gila but not yet about deer as the limiting factor for forage.

The Forest Service, however, was determined to avoid another Kaibab on the Gila.[15] They conducted an exten-

13. "Pineries and Deer on the Gila," *New Mexico Conservationist*, 1:3 (March 1928), 3.

14. It is also curious that Leopold should have made no mention of the role of sheep in the spread of pine and juniper thickets, even though he had hunted at least in part on his brother-in-law's sheep allotment. In the 1920s, to be sure, Ed Otero was grazing his ranges so lightly that Forest Service inspectors repeatedly expressed alarm about the greatly augmented fire hazard. But back in the late nineteenth century the Lunas were reputed to have run 60,000 sheep annually on the Gila; and Leopold would certainly have been familiar with research indicating that browsing by sheep on yellow pine seedlings could virtually preclude reproduction. Reduction of sheep would thus have been an important factor in pine reproduction. Sheep also contributed to the spread of junipers, by depositing seeds in their droppings.

15. Kill figures for the 1927 season are apparently unavailable, but the Black Canyon refuge was opened again in 1928,

sive grazing reconnaissance in 1928, which indicated that a large part of the overgrazing on two central Gila ranges, the Diamond Bar and the GOS, could be ascribed to the increase in deer. As a result of their study, the forest supervisor and district range management officials worked out a tentative plan for drastically reducing the deer population. They estimated there were 21,239 deer on the two ranges. Allowing for natural increase, this ingeniously contrived and carefully calculated proposal provided for removal of 30,381 deer over a period of five years—10,400 by hunting (figuring about 1,000+ hunters per year, about the maximum that could conceivably reach so inaccessible an area, and two deer per hunter), and 19,981 to be disposed of by the state under some plan later to be devised—to arrive ultimately at the 7,834 deer it was estimated the ranges could carry. On the Diamond Bar allotment, on which the Black Canyon refuge was located and which was considered to be understocked with cattle, the full 2,185 head of permitted cattle were provided for under the plan, but on the GOS it was thought that in order to leave sufficient deer to meet the demands of future hunters it would be necessary to reduce the number of cattle from 4,800 to 2,050 head. Even if all cattle were removed from the range, the report noted, there would not be enough forage to carry the present number of deer.[16]

It is hardly surprising that the phenomenon of over-

---

and well over twice as many bucks were shot that year as in 1926. Over 2,000 hunters found their way into the Gila (the entire forest), largely as a result of increased publicity given the Black Canyon area, and they killed nearly 1,000 deer, apparently mostly blacktails. But still the deer increased. See Scotty (J. A. Scott) to AL, 5 March 1929, bound with "Deer Management in the Southwest," LP 6B10.

16. James A. Scott, Memorandum G-Supervision, Gila, 19 Dec. 1928, FRC 316615. The figures in this paragraph were derived by adding Scott's individual figures for the two ranges. Other documents indicate that the GOS outfit was in process of being sold at the time, a circumstance that may have made the proposed reduction of GOS cattle relatively painless. The new owners began restocking the range in 1931, even though the deer problem was not yet solved.

grazing by deer should have been early noted and greeted with alarm by range management officials, in view of the special effort they had been making all during the 1920s to reduce the number of head of livestock and promote range recovery on the Gila and other national forests. They were committed to maximizing the economic utilization of the forests by local interests, such as stockmen, while maintaining the resource base, and they realized that if the deer population were allowed to multiply and the forage situation became critical, the public would demand, as on the Kaibab, that livestock, not deer, be removed. It is perhaps revealing of Forest Service priorities that game work, since its inception around 1915, had been assigned to the office of grazing or range management.

The forest officers' plan for reducing deer on the Gila was put forth just a month after the U. S. Supreme Court handed down its decision in the Kaibab case, upholding the district court as to the government's right to protect its property but with the limitation that the federal government could not license hunters to kill deer in violation of state laws. *Hunt* v. *U.S.* was a landmark case in the definition of federal property rights, but it left the question of jurisdiction over wildlife as unsettled as ever. Even as the Gila plan was issued, a crew of federal employees was on the Kaibab slaying deer, and in ten days killed more than a thousand animals. This sort of official slaughter, however, fueled violent public indignation, and it would not again be resorted to on the Kaibab. Nor, after the Kaibab experience, could it realistically be contemplated for the Gila. As it happened, a new governor, a new game code, and a new game commission in Arizona in 1929 gave at least the possibility of more liberal seasons and bag limits on the Kaibab in succeeding years. But in New Mexico, which throughout the decade had proved more advanced than Arizona in game management affairs, a bill to extend the game commission's authority to lengthen seasons, increase bag limits, and take other measures for reduction of excess deer was introduced in the 1929 legislature only to be, in District Forester Frank

C. W. Pooler's words, "batted about, amended, killed, resuscitated, and finally put to rest." There would be no herd reduction in New Mexico that year.[17]

Deer, Wolves, Wilderness, and Roads

Political maneuvering and factionalism grew more intense, and conditions on the Gila continued to worsen. Leopold, with his wide-ranging contacts in the Southwest, heard accounts of the situation from several angles, but he was not one to be stampeded by political emotionalism. He was, moreover, still engaged in the sober task of determining normal stocking and productivity ratios for use in the second draft of his manuscript on the Southwest. Thus, as a careful student of his own teachings, he kept his eye on the statistics and ratios, watching for what they might tell him. In March 1929 he wrote to Pooler that he was inclined to accept the findings of the most recent Forest Service report indicating severe overstocking of deer in the Black Canyon, particularly since the figures showed a decreasing ratio of bucks to does on the Gila since 1925. Although he could not decipher the meaning of the decreasing ratio, he knew that a similarly excessive number of does was being noticed in Pennsylvania, another severely overstocked area. It had occurred to him, he told Pooler, that this shift in the ratio might not necessarily be due to the overkilling of bucks, as usually supposed, but rather to an accumulation of barren does.[18]

For years Leopold had been worrying about barren does. Barren does meant a shortage of fawns and hence lower deer populations. His changing views on the causes and remedies for "barrenness" provide a lesson in interpretation and rationalization. In 1920 he had thought barrenness was due to buck shortage and, arguing against a doe season, advocated refuges to protect the bucks. Ac-

17. Frank C. W. Pooler to AL, 15 March 1929, LP 3B8. From other friends in the state, Leopold learned that the New Mexico bill had fallen afoul of factionalism, not only in state government but also within the Game Protective Association itself.

18. AL to Pooler, 21 March 1929, LP 3B8.

cording to his theory of refuges the excess bucks produced on the refuges would be crowded off into the surrounding territory, where they would not only redress the buck shortage but also breed heretofore barren does. This remedy as to bucks appeared to be borne out by the hunter censuses conducted on the Gila from 1923 to 1925, which indicated that the proportion of bucks to does increased from 1:4 in 1923 to 1:2.5 in 1925. But still Leopold could not get away from the bane of barrenness. In "Southwestern Game Fields" he minimized buck shortage as its cause and suggested possible physiological factors in the does, including temporary conditions that might be induced by forage of poor quality.

Now in his letter to Pooler, after receipt of figures indicating that the buck:doe ratio on the Gila had fallen back steadily since 1925 to less than 1:4, he introduced a new factor, speculating that the accumulation of barren does might have resulted from the killing off of too many of the large predators. Since the remaining small predators (coyotes and bobcats) preyed largely on fawns and yearlings and hunters were allowed to shoot only bucks, there was no method to remove the superannuated does that were thus overgrazing the range. "I would not like to have this theory given wide publicity," he was quick to caution Pooler, "because it might lead to a demand for an open season on does." It was not clear in Leopold's mind that a doe season was the real remedy. Instead, he suggested, "Concentrating the predatory animal work on coyotes and letting the lions alone for awhile might be a better remedy." This remedy as to does, of course, betrayed his lingering attachment to the buck law, the classic form of hunting restriction.

If the succession of events had not yet led Leopold to change his thinking concerning the advisability of hunting does, it had significantly influenced his thinking on predators since the day in 1920 when he had warned: "It is going to take patience and money to catch the last wolf or lion in New Mexico. But the last one must be caught before the job can be called fully successful." After a trip on the Wichita National Forest in Oklahoma in

1925, he had prepared a memorandum stressing the potential for ecological studies, in which he observed that to facilitate research "it is important to avoid the extermination of predators, but there is no danger of this as yet."[19] The Wichita memo may have been his earliest statement that predatory species, for ecological reasons, ought not to be exterminated. Arguing against the bounty system in "Southwestern Game Fields," Leopold had come out strongly in favor of the system of professional predator hunters which had proven its effectiveness on the Gila, but he no longer spoke in terms of eradication. In fact, he even suggested that ample yarding coverts for deer might reduce the depredations of wolves and coyotes. From this point to the suggestion that predators actually be allowed to trim some of the superannuated does was a short step, relative to the distance Leopold had already traveled. He had not yet given up the idea that control of predators was essential for restoration of deer. Letting mountain lions take a few of the old does seemed safer than opening the hunting season for does, and concentrating control work on coyotes would save a good many fawns.

Leopold's shift of emphasis from control of mountain lions to control of coyotes was in part a response to changing conditions in the game fields, as J. Stokley Ligon was to make clear in his section on predators prepared for "Deer Management in the Southwest." Ligon noted that the predator problem as it had existed some twenty years previous in the southwestern mountains had been solved. The timber wolf had been practically exterminated, and lions and bobcats were held down to low and ever-diminishing numbers. In their place was a newcomer, the coyote, who for some inexplicable reason had recently extended his range into the mountains, where he was an extraordinarily effective predator on fawns, turkeys, and other game. "The coyote," Ligon wrote, "once the familiar clown of the prairie, where it had an economic value as a scavenger and as a check on rodent pests, in its new environment has developed into the predatory animal

19. Memorandum for District Forester Kelly, 22 Nov. 1925, LP 3B9.

menace of North America.... This paradox of the wild,"
he labeled the coyote, "this dare to civilization."[20]

In the Gila, on an area of about 1,000 square miles that
included Black Canyon and the Diamond Creeks, Ligon
estimated there had been 60 mature lions operating full
time prior to intensive extermination activities he directed during 1916–1922. These lions alone, he said, had
been killing 2,500–3,000 deer a year, at a time when
there were practically no coyotes on the area and very
little hunting. By 1928 he estimated there were no more
than 20 mature lions on the entire 2,500 square miles of
the Gila, but coyotes—about 300 of them, each killing
about 12 deer annually, mostly fawns—had "more than
absorbed the savings." He was so impressed with the
destructiveness of coyotes that in his sections of the 1929
draft he featured fawn mortality, rather than "the so-
called barren does," as the real obstacle to deer restoration in the Southwest. "So-called," because many people
simply assumed that does sighted without fawns were
barren. "The 'barren' doe fallacy," he wrote, "is probably
ninety per cent predatory animal imposed." Ligon thought
it significant that when mountain lions had been the principal predators there was never such a shortage of young
deer on the Gila as in later years when coyotes became
abundant.[21]

Although Leopold subscribed to Ligon's interpretation
of the coyote situation, he had by this time probably advanced beyond Ligon and most of his contemporaries in
the development of an ecological attitude toward predators. A year or so after Leopold's warning against the extermination of predators on the Wichita, for example,

20. J. Stokley Ligon, "Predatory Animals and Deer," typescript, c. 1929, 9 pp., LP 6B10.
21. Ligon, "The Unit Herd," and "Predatory Animals and
Deer"; Ligon to AL, 27 June 1929; all in LP 6B10. Ligon distinguished between barren does, which had never conceived,
and dry does, which had lost their fawns through abortion or
shortly after birth. Although he recognized poor forage and
water conditions as factors in abortion and in the death of
young fawns, especially in arid lands where livestock grazed
year round, he regarded predation as the great obstacle everywhere.

Ligon could still write in *Wildlife of New Mexico* that "the complete control [eradication] of predatory animals in the Gila drainage would prove to be of much scientific and economic value" (p. 52). At the same time as Leopold suggested to Pooler that lions be allowed to increase on the Gila, moreover, predator removals on the Kaibab were reaching an all-time high. Behind the concerted federal effort to provide for removal of surplus deer on the Kaibab by hunting was a conviction on the part of many officials that the only other alternative was toleration of large predators, which to their minds was intolerable. The popular though tenuous explanation of later years that the Kaibab problem had been caused by the removal of predators was scarcely heard in the 1920s. Even where excess deer and severe depletion of forage were most clearly evident, there were few who would admit that predators might have any value at all.[22]

By spring 1929, with a problem on the Gila potentially as severe as on the Kaibab and with a number of options for dealing with it apparently foreclosed by legislatures, courts, public opinion, or trial and failure on the Kaibab, the Forest Service appointed a special investigative committee to examine the deer situation on the GOS and Diamond Bar ranges and to make recommendations. The group was chaired by C. E. Rachford, chief of range management in the Washington office, and included Walter P. Taylor of the U. S. Biological Survey, who had earlier examined the Kaibab and who was to work with Leopold on problems of excess deer for two decades to come, as well as other representatives of the Survey, the Forest Service, the state game department, and the Silver City

22. See for example E. A. Goldman, "Surplus Game: A Problem in Administration," *American Game* (April 1926): "Excessive numbers of game may be kept in check by tolerating the presence of large predatory animals that prey upon them, but human interests are paramount, and they are best subserved if the predatory animals are kept under rigid control." For a critique of the notion that the Kaibab problem was caused by removal of predators see J. Burton Lauckhart, "Predator Management," 41st Annual Conf., Western Assoc. of State Game and Fish Commissioners, New Mexico, 1961, pp. 85–88.

game protective association. Not unexpectedly, their investigation of conditions corroborated Ligon's report of 1927 and the Service's own range reconnaissance of the previous fall.[23]

About half of the committee's recommendations concerned long-range policy and, not surprisingly, in view of the number of committee members who had been associated with Aldo Leopold for years while he was developing his ideas on forest game management, read like a chapter out of Leopold. The committee recognized the preeminent need for an aggressive campaign for proper management of forest and forage, called for the assignment of a competent biologist to study the Gila herd, and stressed the need for adequate scientific data on food habits and life histories of game and reliable techniques for determining numbers of game animals, rate of increase, sex ratios, and the effect of predators and disease—the very types of data Leopold had been seeking for his deer management manuscript and for lack of which he was floundering.

The other half of the recommendations dealt with specific measures for immediate reduction of the herd. These recommendations, although basically sound, were strung along a single strand of logic, the end point of which Aldo Leopold would have had difficulty accepting. Immediate relief was desirable. It could best be secured by means of hunting, including hunting in the Black Canyon and North Sapello game refuges, a longer season, two or more deer per hunter, and killing of does as well as bucks. The only measure that was permitted under state law was the opening of refuges, the bill to extend the authority of the state game commission having failed that winter. Partial

23. Memorandum on Examination of Deer Range on Gila National Forest (May 1–6, 1929), and associated correspondence, Federal Records Center, Denver, Container 316615 (hereafter cited as FRC 316615). The New Mexico Game Protective Association did not participate in the investigation, although it had received at least three invitations. Pettit asked the Forest Service to present the findings at a meeting of the Albuquerque GPA, but the group did not appear disposed to cooperate in the solution of any problems.

relief, therefore, could be secured only by increasing the number of hunters and thereby the kill of bucks, and by allowing an increase in predators, which could take does and fawns as well. The group agreed that control of lions on the GOS and Diamond Bar ranges should be abandoned temporarily but that the Biological Survey should continue control of coyotes and bobcats, "to afford protection to turkeys."

Finally, to increase the number of hunters, the area would have to be made more accessible by roads. The committee accordingly suggested reconstruction of the old North Star and Copperas Canyon–Gila Flats roads, which were actually wagon tracks used by the army during the Apache War of the nineteenth century. Most of the area involved was within the Gila Wilderness, where roads were ostensibly prohibited. It was, however, the judgment of members of the party that the Forest Service ought seriously to reconsider the need for wilderness classification of that portion of the area lying east of the Gila River (about one-third of the whole) and, alternatively, the effect of opening the area to hunting and the value of roads for fire protection.

If Leopold had been on the scene in New Mexico, he would undoubtedly have been more concerned about the necessity for drastic reduction of deer than he appeared to be from his vantage point in Wisconsin. No one was more concerned about overgrazing and watershed damage than he. He himself had suggested abandonment of lion control in his letter to Pooler that March. He would probably also have favored the opening of the refuges and acknowledged the desirability of more hunters and more liberal regulations, including the killing of does. But he would have been torn by the thought of roads in his wilderness.

"Who wants to stalk his buck to the music of a motor?" Leopold had asked years before, "Or track his turkey on the trail of the knobby tread?"

> Who that is called to the high hills for a real *pasear*
> wants to wrangle his packs along a gravelled highway? Yet
> that is what we are headed for, at least in the Southwest.

Car sign in every canyon, car dust on every bush, a parking
ground at every waterhole, and Fords on a thousand
hills! [24]

At the time the Gila was designated as wilderness, the
only other large roadless area remaining in the Southwest
was on the Kaibab National Forest (west of Highway 67
to the north rim of Grand Canyon). But the Kaibab could
not remain roadless, Leopold had pointed out, and the
reason ironically was the need to eliminate excess deer:

> The Kaibab is a country of limited water-holes. Appar-
> rently it will require somewhere around 2,000 hunters
> each year to utilize the natural increase of the deer (would
> this were true elsewhere!). To camp 2,000 hunters on a
> couple of dozen water-holes would entirely destroy the
> recreational value of the hunting. These hunters will have
> to camp dry; dry camps are practicable on such a scale only
> with motors; and motor camps mean at least some roads. [25]

Since it had first been proposed that the Gila be designated as wilderness, the idea had been opposed principally by forest officers and private interests who wanted a
road along the old North Star route from Mimbres to
Beaverhead—to get at inaccessible timber, to serve private
holdings, to shorten the distance from Silver City to Magdalena, and for various other purposes, often advanced
under the guise of fire protection. Aldo Leopold, who
was in charge of fire protection for the entire Southwestern District in those days and in 1922 had developed a
highly regarded "sample fire plan" for the Gila Forest,
was of the opinion that roads in that area were not needed; indeed, that "more roads would add as much or more
fire risk than they would help combat." But pressures for
a road bisecting the wilderness continued unabated. In
1927 Forest Service officials held a special conference on
the subject and decided "that nothing would be done
toward the improvement or extension of the road . . .

24. "The Next Move in Public Shooting Grounds," holograph, c. 1921?, 1 p., LP 6B16.
25. "A Plea for Wilderness Hunting Grounds," *Outdoor
Life*, 56:5 (Nov. 1925), 349.

until through the actual occurrence of fires or through some unforeseen contingency it was demonstrated that the road would be needed."[26]

The "unforeseen contingency" turned out to be an irruption of deer in the very center of the region through which the old North Star route passed. Against a contingency like this there could be little argument, for overbrowsing by deer was a threat to the integrity of the wilderness itself and would have been recognized as such most clearly by the very men who had heretofore been the most stalwart defenders of the wilderness idea against merely private or economic interests. District Forester Pooler, who in June 1924 had designated the area a wilderness, in June 1929 made the decision to reconstruct the road.

Although the intent was to have the North Star Road passable by automobile for the 1929 hunting season, the road was not actually completed until 1931. In April of that year, Leopold and Fred Winn, who had been supervisor of the Gila when it was designated wilderness, drafted and circulated a petition proposing that because the road was wholly for protection purposes, the Forest Service should keep it closed except when it was actually being used for protection. Both men knew the wilderness intimately and knew what the road foreboded; but their petition was to no avail.[27]

The North Star Road invaded what some call "the soft country" of the Gila—the open, rolling, mesa-and-canyon terrain where wild game abounded and the aura of the old West lingered on. Others, in later years, would claim that such country was not "true wilderness," in the sense of rugged mountains like the Mogollons to the west. But to Leopold, true wilderness meant an integral watershed, encompassing diverse terrain and even packsaddle stock

26. Leopold, "Report on Proposed Wilderness Area"; Hugh G. Calkins, Memo for Files, 30 March 1927; both at Gila Forest headquarters, Silver City.

27. AL to Ward Shepard, 6 April 1931, LP 3B3. "I am fully convinced," Leopold wrote, "that the heavy stocking of deer following upon the pre-forest overgrazing by cattle has produced a serious situation. Whether this could have been corrected without a road, or whether the road will correct it, are two questions to which I do not know the answer."

ranches, protected from motorized access by mountain ranges and box canyons. It was the soft country in the interior, not the rugged barrier mountains, that contained most of the game. The North Star Road, in bisecting the wilderness, led to separate administration for the barrier range to the east, designated the Black Range Primitive Area. More important, the road made it possible for deer hunters and others to strike off across the relatively easy terrain to the west almost anywhere they wished and penetrate many miles into the wilderness in their motors. Within a few years, the second recommended road up Copperas Canyon was reconstructed, and the area between them—including much of the Diamond Bar and GOS ranges, the richest deer country on the Gila—was so thoroughly violated that the Forest Service would seek to eliminate it from the wilderness system. The Gila, regarded by many as the prototype of wilderness areas in national forests, has been among the most abused of them all.[28]

In an unpublished autobiographical sketch originally intended as a foreword for the collection of essays that was to become *A Sand County Almanac*, Aldo Leopold candidly expressed what the road had come to signify, by 1947, in his own intellectual development. It was inextricably bound in with wilderness, of course, and with deer management, but also with wolves and mountain lions. He told how he had been "accessory" to the extermination of the lobo wolf from Arizona and New Mexico and how he had been able to rationalize it "by calling it deer management." He went on to describe his role in getting the Gila headwaters withdrawn as wilderness, "to be kept

28. When the Forest Service proposed formal classification of the Gila as wilderness under Regulation U–1 in 1953, they left out the entire area between the North Star and Copperas Canyon roads, as well as a number of other sections around the edges, including choice timber acreage in the vicinity of the N-Bar Ranch. Because of public pressure brought to bear by the Wilderness Society and conservationists across the nation and by New Mexico's Sen. Clinton Anderson, a compromise was reached by which the excluded tracts were to be retained in "primitive area" status. As such, they did not receive immediate protection under the National Wilderness Preservation Act of 1964 but were to come up for review within

as pack country, free from additional roads, 'forever.' " But the deer herd, "by then wolfless and all but lionless," he wrote, soon multiplied "beyond all reason" to the point where reduction was imperative:

> Here my sin against the wolves caught up with me. The Forest Service, in the name of range conservation, ordered the construction of a new road splitting my wilderness in two, so that hunters might have access to the top-heavy herd. I was helpless, and so was the Wilderness Society. I was hoist of my own petard.[29]

Vagaries of Herd Reduction

Leopold hunted deer in the Gila for the last time in November 1929, going in via Beaverhead, northern terminus of the North Star route, to the Whiterock Tank region some distance northwest of Black Canyon. Again he found deer wonderfully abundant. His brother Carl gunned down an eleven-point buck on opening day; but he and his son Starker, hunting with bows and arrows, missed dozens in the thick brush. Although Leopold was annoyed

---

ten years, along with the Black Range and other primitive areas, for classification under the act or elimination from the system.

In 1971 the Forest Service, to honor Leopold, proposed establishment of a 188,179-acre Aldo Leopold Wilderness in the Black Range, the least representative portion of the original area. In 1972 they proposed adding some 70,000 acres of the Gila Primitive Area to the Gila Wilderness but declassifying the remainder of the primitive area, over 65,000 acres, most of it in the soft country between the two roads. As of June 1974 the proposals were in abeyance because of a flurry of new mining schemes in the Black Range and a flood of applications for geothermal leases in the Gila Primitive Area.

For an excellent administrative history of the Gila Wilderness see Richard E. Warner, "The Dismembered Wilderness," *Pacific Discovery*, 15:4 (July–August 1962), 2–10. Warner gives fire control and general administrative convenience as justifications for the North Star Road and does not mention the deer situation.

29. "Foreword," typescript, 31 July 1947, 9 pp., LP 6B17. The reference to the Wilderness Society, which was not in fact organized until 1935, is an intriguing lapse of memory.

by all the hunters whom he encountered practically every day, in contrast to previous trips when he had seen other parties only once or twice all told, the number was still insufficient to adequately reduce the deer population.

By this time, Leopold had probably abandoned his manuscript on deer management, but he had not lost interest in southwestern conditions. His concern was undoubtedly stimulated by a report on the Kaibab by H. L. Shantz, botanist–president of the University of Arizona, to the effect that it would require from ten to fifty years to restore the damage excess deer had inflicted on the vegetation of that forest. In the months after his return from the Gila, he published a number of articles on forest-game management, among them "Environmental Controls for Game Through Modified Silviculture" and "Game Management on the National Forests," in which he contrasted the Kaibab, doomed for decades to come, with the Gila, where environmental controls might yet avert disaster.

On the Gila, he explained, yellow pine was growing up in thickets, thrifty junipers were invading miles of country formerly treeless, and grass in some areas was showing satisfactory recovery; but the bread-and-butter browse species, live oak and mountain mahogany, were being eliminated, thereby reducing the future carrying capacity of the range for deer. Part of the browsing was done by deer, he admitted, although the original damage was probably inflicted by cattle. Foresters had reason to be pleased with the successful restoration of conifers and grass; but it seemed to him that in the Gila wilderness with its high recreational values, creation of a balanced environment for game, as well as for timber and livestock production, ought to be a primary objective of Forest Service policy. Leopold admitted he did not know how to stimulate browse reproduction (in a talk at the University of New Mexico in 1933 he actually suggested fencing south-facing slopes to let the mahogany come back), but he had unbounded faith in the ability of Forest Service research to develop a solution, if only the matter were given sufficient priority. It is worth noting that in these articles he still did not recognize the relationship between the browse problem and the gradual encroachment of thick stands of pine and juniper in the absence of fire; nor did he connect the elimination of live oak and mahogany directly to an over-population of deer, or indicate that deer numbers might have to be reduced for a time in order to allow silvicultural manipulation to restore food supplies.[30]

The dimensions of the Gila problem were much larger than anyone including Leopold had yet realized. Early in 1930 the Forest Service secured the services of M. E. Musgrave, formerly with the Biological Survey as Arizona state leader of predatory animal control, and assigned him full time as game specialist for District 3. Musgrave rode over an area of more than three-quarters of a million acres, parts of which he reported were already in as bad condition as the Kaibab, while the total area and numbers of deer were greater than the Kaibab. The range would be utterly destroyed so far as deer and turkeys were concerned unless drastic measures were taken over the entire area in the very near future. But the law still provided for only one buck per hunter, and the New Mexico legislature would not be back in session for another year.[31]

30. "Environmental Controls for Game Through Modified Silviculture," *Journal of Forestry*, 28:3 (March 1930), 321–26; "Game Management in the National Forests," *American Forests*, 36:7 (July 1930), 412–14. See also "Ecology as an Applied Science," typescript, 5 May 1933, 4 pp., LP 6B14.

31. M. E. Musgrave, Memorandum for District Forester, 3 May 1930, FRC 316615.

The official Forest Service estimate of deer on the Gila in 1929 was 44,000, as compared with its estimate of 30,000 when the Kaibab herd peaked in 1924, and as compared with Leopold's estimate of some 4,000 deer on the Gila in 1916 and Ligon's estimate of 41,000 for the entire state of New Mexico as of 1926.

After his examination of the Gila, Musgrave spent a month riding over another range overpopulated with deer, Leopold's old stamping ground in the Blue Range of the Apache Forest, which was "protected" by the 600,000-acre Blue Range Game Preserve. He found heavy overbrowsing especially in the vicinity of Blue River, with mahogany, cliffrose, live oak, juniper, and other palatable species taking on the same aspect as on the Kaibab and in Black Canyon. See M. E. Musgrave, "Report of Game Situation in Blue Range Game Preserve," typescript, 12 July 1930, 6 pp., FRC 210433.

While New Mexico marked time, Arizona advanced to the reduction-and-reaction phase of the deer problem. The newly established state game commission provided for two long two-deer seasons on the Kaibab in 1929 and 1930, in which a total of nearly nine thousand deer were taken, most of them by out-of-state hunters. Public reaction was tempestuous; and Governor Hunt, who had returned to office and was violently opposed to the Kaibab deer program and to the Forest Service, tried to get the legislature to abolish the fledgling commission, which in his opinion had surrendered Arizona's state rights. Leopold, reaffirming his own faith in commissions as a permanent principle of administration, encouraged a number of national conservation leaders to write endorsements of the Arizona commission.[32] The commission was not dissolved, but the combination of politics and volatile public opinion guaranteed that on into the indefinite future the Kaibab situation would remain precarious.

In 1931 the New Mexico legislature finally enacted a bill that transferred authority over seasons and bag limits to the state game commission. Now, with herd reduction a distinct possibility and public opposition to the idea already beginning to mount, the Forest Service fielded another joint investigative committee to examine the Gila ranges. The committee, composed of ranchers, sportsmen, and state and federal officials, recommended opening three refuges to hunting but favored a two-deer season only in the immediate vicinity of Black Canyon, an area of only about 100 square miles, instead of an 840-square-mile area that Musgrave and the Forest Service believed was in need of immediate reduction. They also reversed the 1929 recommendation concerning predators by vigorously opposing any slackening of efforts to control them, so as not to "deprive the sportsman of his just prerogative of killing available game for sport."[33]

32. See correspondence with Fred Winn, Seth Gordon, and P. K. Whipple, January–March 1931, LP 3B3.

33. "Report of Committee that Investigated the Big Game Ranges on Portions of the Datil and Gila National Forests," typescript, 17 July 1931, 5 pp., FRC 316615.

The package of recommendations was clearly a compromise, and a number of participants in the investigation were so dissatisfied that they felt compelled to make their own supplemental reports. Two such reports were written by game specialist J. Stokley Ligon of the state game department and game specialist M. E. Musgrave of the Forest Service, both of whom had formerly been with the Biological Survey as predator control inspectors. Their reports illustrate the difference in interpretation that was possible between individuals with similar backgrounds in the employ of two different agencies, and they foreshadow a divergence in approach and objectives that would widen with the years.

Ligon, whose original report on the Gila deer situation in 1927 had indicated severe overpopulation not only in Black Canyon but also in the entire upper Gila drainage, was now impressed with what seemed to him "a marked scarcity of deer" everywhere except in Black Canyon. There were 50 per cent fewer deer in the drainage than five years before, he judged, and gave it as his opinion that the coyote was largely responsible. "Unless the campaign against predatory animals is vigorously prosecuted throughout this region," he warned, "a closed season on both deer and turkey in order to save breeding stock is impending." By the same token, he opposed any liberalization of hunting restrictions outside the immediate Black Canyon area. Although he saw evidence of excessive browsing everywhere, he contended that this abuse had its inception several years previous and that the browse, except for the young junipers, was now showing satisfactory recovery. Summing up, he wrote:

> Some of the Forestry officials on the committee constantly expressed alarm regarding forage conditions, while with myself, as well as others including the Game Warden, alarm had to do more with scarcity of deer. The natural conclusion is, assuming there are fewer deer than formerly as well as fewer cattle, the necessary remedy is already being applied to insure forage restoration.[34]

34. J. Stokley Ligon, "Special Report on Investigation of Big Game Ranges on Portions of the Gila and Datil Forests,

Musgrave conceded that in certain areas that had been heavily overutilized in the past, such as the Diamond Creeks, deer were less abundant than formerly. However while Ligon had blamed this scarcity of deer on predation, Musgrave believed that deer were being forced out because of continuing scarcity of forage and were drifting south and west, where they were compounding the problem over an increasingly large area. He explained that some members of the committee who thought the forage was showing satisfactory recovery had been misled by the ideal growing conditions that spring. He was concerned with soil erosion attributable to localized overbrowsing by deer, especially along canyon bottoms, which had recently become so heavily washed out that it was practically impossible to ride a horse across them, and on many south-facing slopes. "The heavy concentration of deer on these south slopes," he explained, "where the snow melts off first and where the more palatable plants grow is causing the sloughing off of the soil which is having its everlasting effect on these ranges." He recommended a two-deer season for virtually the entire area. As to predators, Musgrave simply made no mention.[35]

Both men were concerned about the scarcity of turkeys. Ligon thought it due to predation and Musgrave to destruction of food and cover by deer. Regarding damage to yellow pine reproduction, Musgrave and other Forest Service men tended to blame deer, while game department officials at one point on the trip jubilantly spotted cattle nipping tops from young pines—an indication, Ligon

---

July 1931," typescript, 5 pp., Ligon Papers, Conservation Library Center, Denver Public Library.

As for the acknowledged overbrowsing of young juniper, Ligon was not particularly concerned: "Even though deer may kill some of the reproduction, we may ultimately find that this elimination is not undesirable in view of the rapidity with which the new growth is taking valuable grass lands." He was one of the first to make this point; but he apparently had not yet recognized the corollary, that the juniper would inhibit not only grass for livestock but also other forage plants required by deer.

35. M. E. Musgrave, Memorandum for Regional Forester, 14 Sept. 1931, FRC 316615.

suggested, that cattle might be responsible for considerable damage of other sorts attributed by the Forest Service
to deer.

Such differences of opinion among technically competent men, bearing on long-range policy, obviously could
not be aired before the general public, who distrusted even
the idea of reducing deer in the limited area of Black
Canyon. Tensions were already flared that summer of
1931. Instead of continuing the three duly appointed
game commissioners in office, the governor asked for res-
ignations and appointed a new slate, which in turn fired
the state game warden and appointed a different man,
Leopold's associate from long ago on the Carson, Elliott
Barker. The ousted warden, E. L. Perry, became spokesman for a faction in the state game protective association
that opposed herd reduction and regarded the game commission's new grant of authority over seasons and bag
limits as three small votes from disaster. Joined with Perry
was R. Fred Pettit of the erstwhile Leopold, Ligon, and
Pettit deer management manuscript.

In the midst of this dissension appeared a lone issue of
the *Pine Cone,* the newspaper Leopold had founded in
1915 to promote the cause of game protection and creation
of a nonpolitical commission for New Mexico. Leopold
himself had brought out the first seventeen issues, up to
enactment of the commission law in 1921, and apparently
also the eighteenth, issued in March 1924 just before he
left New Mexico. The nineteenth and last issue was
brought out in July 1931 by E. L. Perry. Perry paid glowing tribute to Leopold in a front-page history of the gold-
en days of the game movement in New Mexico, then
caustically criticized the current political regime in the
state, striking a traditional sportsman's stance:

> The G. P. A. has one basic conviction, and that is that the
> Game Commission owes its allegiance to the sportsmen, and
> not to any political party, personality, or set of prejudices.
> The sportsmen contribute every dollar which the Commis-
> sion has to spend, and theirs must be the guiding voice in the
> conduct of the Game Department.

(Leopold may have at one time believed that a game commission owed its allegiance to the sportsmen, but by 1931, after having completed his game survey of the north-central states, he was convinced that the obligation engendered by financing state game departments exclusively from the sale of sporting licenses was one of the biggest impediments to a balanced conservation program. The situation in New Mexico might have strengthened that conviction.) After the G. P. A. convention later that summer, Perry wrote Leopold that the association had just been taken over by a rival faction; it would henceforth be "a submissive adjunct" of the game commission, and the *Pine Cone* would be discontinued. Leopold's seemingly detached reply: "I am afraid that game movements, like grouse, are cyclic."[36]

For Leopold the news from New Mexico must have been painfully disturbing and confusing. His closest friends from his years in the Southwest were now spread over at least three different, diverging camps. In the Forest Service calling for massive deer herd reductions on the Gila were his longtime associate, District Forester Frank Pooler, along with Winn, Scott, Locke, Musgrave, and others. Willing to acquiesce in one liberal season in the immediate vicinity of Black Canyon but increasingly distrustful of the Forest Service's larger designs were Elliott Barker, originally a Forest Service man himself, and Leopold's former co-author, J. Stokley Ligon, both now in the state game department. Disillusioned with both the Forest Service and the game department and watching the New Mexico Game Protective Association crumble around them were his stalwart from the earliest years of the game protection campaign, R. Fred Pettit, and the former state warden, Edgar Perry. It is hardly surprising that Leopold himself, fifteen hundred miles away and with better bonds of friendship than sources of objective information, refrained from taking sides.

Acting on the compromise recommendation of the joint investigative committee, the game commission of New Mexico established a two-deer (does included) season for

36. AL to Perry, 14 Sept. 1931, LP 3B8.

a 100-square-mile area in the vicinity of Black Canyon. Some 2,440 hunters traveled the newly opened North Star Road to the two-deer area and took 2,335 deer—a kill of about twenty-three deer per section, not counting animals that were killed and left in the woods. Forest Service officials were delighted with the reduction and looked forward to liberalized seasons over more of the overbrowsed Gila ranges in succeeding years.[37]

But public reaction was immediate and adverse, and it was directed primarily at the state game department, which had managed the hunt. From the "Public Forum" of the *New Mexico State Tribune* (Albuquerque), for example:

> One of the finest deer areas of the state has been laid waste. . . . Even the little fawns, the baby deer that might have furnished the nucleus for new herds, have been wantonly slaughtered by an army of hunters who invaded the open area and blazed away at everything that had four legs. . . . Not sportsmanship—murder.
>
> Rotten partisan politics has again grasped the Game Department of the state in its talons, setting at naught the endeavours of the State Game Protective Association and other game protective bodies to protect and perpetuate game and wildlife in the state. . . .
>
> The State Game Department is directly responsible for one of the most horrible orgies of killing that was ever perpetuated in the name of sportsmanship. "Not sport—but slaughter."[38]

In an effort to quell such criticism, Ligon wrote an article justifying the Black Canyon two-deer season and asserting a unanimity of opinion among game department and Forest Service officials. From long scarcity of

37. "The Black Canyon Deer Hunt," 14 Nov. 1931 (news release, Forest Service, Albuquerque), FRC 38301. Of the total of 2,335 deer, 1,684 were does (except for a few spike bucks), 601 were bucks, and 50 fawns.

38. Excerpted from an article by Larry Bynon, 26 Oct. 1931, quoted in Robert H. Stewart, "Historical Background of the Black Range," 30 July 1962, W–75–R–9, Work Plan 20, Job 7 Completion Report, New Mexico Department of Game and Fish, pp. 24–25.

suitable food and perhaps because of forced in-breeding on congested range, he explained, there had resulted physical abnormality of the herd as a whole. Hence it had been necessary to virtually eliminate the herd rather than merely reduce it in order to allow for the recovery of forage and the reestablishment of a normal, healthy herd:

> We had here, we now realize, a situation probably never before met with in game management: a degenerated herd, under-sized, under-developed and characterized by poor flesh, freak individuals and serious shortage in reproduction, probably one doe out of three showing evidence of having fawns.

(This explanation for the poor fawn crop was, of course, quite different from the one he had offered that summer blaming predators. In a game department report in December he made his shift in explanations explicit: "This year the short fawn crop could not be attributed to coyotes or other predators.") In an orgy of apologetics, Ligon continued:

> To the average observer who visited the area there was no serious over-abundance of deer nor shortage of food; but to capable, free-thinking and free-acting men of the Forest Service and Game Department, who were closely associated with the situation and who have only the best interest in the future welfare of our natural resources at heart, the situation represented a different picture.

Ligon's justification referred only to Black Canyon, yet the difference in tone, as compared with his report on the deer situation that summer, is striking and may be taken as an indication of the intensity of public reaction that the state game department and the Forest Service together were confronting.[39]

39. J. Stokley Ligon, "Black Canyon Deer Removal Justified" (typescript copy of article in Silver City *Independent*, 10 Nov. 1931), FRC 316615; game department report quoted in Stewart, "Historical Background of the Black Range," p. 16. Ligon estimated that 85 per cent of the deer had been removed from the area as a whole, and 95 per cent in Black Canyon proper; the foreman of the Diamond Bar ranch, however, es-

R. Fred Pettit, in a letter to Leopold, expressed a critic's view of Black Canyon and of Ligon's apologia:

> The NM Game Dep't is steadily pry into things and fields into which it has no advance knowledge, results are pretty discouraging to the sportsmen; they cleaned out the socalled "Black canyon area" last fall of deer, it is now proposed to do the same in the Guadaloupes, open the season on antelope, and such other bids to the malcontent. . . .
>
> The Game Dep't very adroitly maneuvered Ligon into making their defences, and that ruined good old stalwart Ligon with the boys thruout the state, they felt he had "sold out", it was one of the damndest pieces of jockeying I have witnessed.[40]

Despite or perhaps because of public opposition, individuals in key positions in the Forest Service and the state game department had cooperated adequately for a time to secure a limited objective on which they could agree—reduction of the herd in Black Canyon. But after the dust settled, all the subtle and not so subtle differences in interpretation and objectives reasserted themselves.

The extent of divergence was dramatically revealed in a confidential statement prepared by J. Stokley Ligon for the state game commission in 1934. Ligon, who had praised the "capable, free-thinking and free-acting men of the Forest Service" after the 1931 reduction, now in 1934 was charging that the Forest Service attitude toward big game in New Mexico was "the most serious menace to such game in the state." He had just returned from an inspection of the Black Canyon country; so shocked he was at the little evidence of deer reoccupation of the range that he urged reestablishment of the former Black Canyon refuge, which had been abolished in 1932. But Forest Service policy, he had come to realize, was now especially averse to big-game refuges on the national forests. Big

---

timated the stocking within the two-deer area at 60 per cent of the normal past stocking (G. L. Wang, Memorandum for Supervisor, 20 Feb. 1932, FRC 316615).

40. R. Fred Pettit to AL, 11 March 1932, LP 3B8.

game was being discriminated against, he thought, in favor of grazing by domestic livestock. He charged that Forest Service figures indicating a substantial annual increase of deer on New Mexico forests with only a 5 per cent loss due to predatory animals were at variance with the facts, that it was doubtful whether there were as many deer as there had been four to five years previous, that predators reduced the herd by at least 20 per cent, and furthermore, that the Forest Service might have had an understanding with the Biological Survey to "go light" on predators.[41]

In short, the Forest Service seemed intent on managing deer to decrease the population, while the game department wanted to manage to increase it. At the first sign of deer abundance on the Gila, Aldo Leopold had urged New Mexico sportsmen to concern themselves with increasing the carrying capacity of the ranges, warning that "game conservation without range conservation is an idle dream." Black Canyon taught foresters the converse, that range conservation without game management was just as unthinkable. Overbrowsing by excess deer was imperiling the stockraising business, and the Forest Service felt a responsibility to help maintain the local economy. In the context of crisis and in the absence of any congressional mandate for wildlife restoration on the national forests, game management as developed by the Forest Service came to be essentially a matter of managing downward, of reducing excess deer or preventing their increase, rather than of building up carrying capacity to sustain larger populations in areas that were depleted. Leopold early recognized this tendency. When in 1931 the New Mexico Game Protective Association organized a new Board of Research and Education to furnish a balance wheel for wildlife management in the state, he volunteered the suggestion that their campaign include "some pressure on the Forest Service to actually practice game

41. Ligon, "A Confidential Statement," 18 May 1934, and "Progress in Deer and Forage Restoration at Black Canyon," 26–27 April 1934, Ligon Papers, Conservation Library Center, Denver Public Library.

management on its idle territory as well as on its over-stocked areas."[42]

The game department, by contrast, was still trying to increase deer populations by stricter law enforcement, predator control, and refuges. On brushy ranges with practically no grass, a condition that, as Leopold had early noted, was becoming increasingly common on the Gila and elsewhere in the Southwest, Elliott Barker and other state officials believed that deer and cattle were in competition for 90 per cent or more of the forage and that destruction of browse plants was due more to stock than to deer. Hence they insisted that the Forest Service remove the cattle, which could not be raised profitably on such poor ranges anyway, and turn the areas over to deer, which they considered the highest use.[43] What

42. AL to E. L. Perry, 2 June 1931, LP 3B8. Although Leopold was apparently unaware of it, Black Canyon had provoked some discussion in the Forest Service in 1931 on the question of what constituted satisfactory stocking or overutilization of a range by deer and cattle. The inspector of grazing, D. A. Shoemaker, and others held to traditional Forest Service range policy, favoring toleration of limited overutilization by deer as well as by cattle, while Musgrave, concerned more with erosion and range quality, was inclined to interpret any overutilization as evidence of present or potential overstocking. Regional Forester Pooler, in calling for a full discussion of the varying viewpoints on proper stocking, pointed out that "first indications of local over-grazing cannot be ignored in the case of deer with impunity." The Forest Service, Pooler seemed to be suggesting, had much less control over deer populations than it did over cattle, for political as well as biological reasons. It is not hard to imagine that when the Forest Service called for drastic reductions of deer they were following Pooler's rationale rather than Musgrave's. See D. A. Shoemaker, Memorandum for Regional Forester, 30 Sept. 1931, FRC 3617; M. E. Musgrave, Memorandum for Regional Forester, 14 Sept. 1931, FRC 316615; Frank C. W. Pooler, Memo for Mr. Kerr, 27 Nov. 1931, FRC 38301; and other related memoranda.

43. Elliott Barker, "Report to Game Commission on Inspection of Black Canyon and Adjacent Areas, March 23 to March 31, 1932," FRC 316615. After their initial defensive reaction to Barker's suggestion, the Forest Service gradually began reducing livestock permits as the cattle ranchers left for financial reasons (occasioned in part by the deterioration of the range). During the same years the Luna–Otero–Bergere sheep were be-

neither side appreciated was the extent to which both species would have to be reduced if the ranges were to recover.

In the atmosphere of distrust generated by divergent interpretations and objectives, cooperation between the Forest Service and the state could hardly proceed smoothly. Black Canyon refuge was reestablished by the New Mexico game commission in 1936, as recommended by Ligon "to assist in restoration of normal deer hunting"— this, despite the findings of yet another joint investigation that browse foods, especially live oak and mountain mahogany, were in worse condition than they had been in 1931. The state also stepped up its own predator control operations. But nobody tried to do anything about the browse problem.[44]

When Aldo Leopold made a national survey of overpopulated deer ranges in 1947, he had to list the Black Canyon as a chronically overstocked area. The Kaibab was on his list too, and so were the Magdalena Mountains, the area he had originally selected in 1914 on which to determine normal deer stocking and productivity.[45]

---

ing gradually withdrawn from the Mogollons farther to the west (also for economic reasons, though not because of range deterioration). By the 1960s much of the central core of the Gila Wilderness Area was free of livestock.

44. Stewart, "Historical Background of the Black Range."

45. Aldo Leopold, Lyle K. Sowls, and David L. Spencer, "A Survey of Over-Populated Deer Ranges in the United States," *Journal of Wildlife Management*, 11:2 (April 1947), 162–77. An inspection of deer and range conditions on the Magdalena Division of the Cibola (formerly Datil) National Forest was made by E. A. Schilling, assistant regional forester, in November 1948. He found the area in deplorable condition. "Regretable, indeed," he wrote, "that someone's imagination went so far as to classify such an area as cow range. And that soil erosion of such serious nature is still tolerated, is more regrettable." The Forest Service had been concerned about deer concentrations as early as 1932, but there was no concern then about livestock grazing. In 1934, J. Stokley Ligon had been the first person to recommend that livestock be removed and the area devoted to watershed, game, and recreation. Foresters had insisted on the value of the area as livestock range, but Schilling stated, "In my opinion he [Ligon] spoke the only truth throughout the years." Now, in addition to removing livestock, it was

Studies conducted during the early 1960s by the state game department, under funding from the Pittman-Robertson program, indicated that deer populations in the Black Canyon area had declined about 65 per cent since the late 1920s, and the decline could reach 90 per cent. The decline could be attributed to loss of the staple deer foods, live oak and mountain mahogany, 90 per cent of which was dead or dying. Over the Gila Forest as a whole, the Forest Service estimated fewer than 25,000 deer on 2.7 million acres in 1973, as compared with 44,000 deer on 1.8 million acres in 1929. Neither herd nor range had yet reached equilibrium.[46]

The Deer–Environment Equation

In "Southwestern Game Fields," especially in his chapter "The Virgin Southwest and What the White Man Has Done to It," Leopold had the essential ingredients for an interpretation that could have accounted for the fortunes of deer on the Gila. For a variety of reasons, including not only his lack of experience with deer populations in those days but also his attitudes and values as a forester and his concentration on the technical aspects of game management, he did not successfully integrate the role of deer into his interpretation of environmental change in the Southwest. When the crucial events occurred that could have caused him to reevaluate his whole analysis of the deer situation, he was hundreds of miles away in Wisconsin. He did return for two bow-and-arrow hunts, but those were vacation trips and he stayed away from the worst problem areas. He did not have an opportunity for the sort of concentrated field experience that might have led

---

going to be necessary to reduce the deer to a negligible quantity and, of course, for the state game department to cease predator control operations. "Browse, soil, ground cover, is terribly sick," Schilling concluded, "and the land, if it is worth anything, is deserving of a long rest." (W–Inspection–Cibola, 6 Dec. 1948, FRC 38301)

46. The Pittman–Robertson studies are summarized in Stewart, "Historical Background of the Black Range." The 1973 estimate is from Joe Janes (Forest Naturalist), letter to author 28 June 1974.

him to rethink the whole interpretational structure he had built up during his previous years of observation in the national forests of the Southwest.

If he had been on the scene at the time, or if he had reflected on the deer–environment equation in the Southwest from the perspective of his experiences later in life, his interpretation might have gone like this: Overgrazing, exclusion of fire, and probably also climatic changes allowed enormous quantities of brush to "take the country" along the Mogollon Rim. Deer, formerly held to low levels by a combination of predatory animals, Indians, settlers, and market hunters, began to show an increased breeding rate and a greater ability to escape or hide from predators and hunters under the excellent conditions of food and cover provided by the brush fields; the growing deer population was further protected by the confining of Indians to reservations, prohibition of market hunting, the one-buck law, strict enforcement, predator eradication, and refuges. As areas that had formerly been grassy were overtaken by brush and by thick stands of young pine and juniper, cattle were forced increasingly into competition with deer for palatable browse; and the combined pressure of cattle and deer depleted available browse supplies and tended to favor the less palatable species. New reproduction of browse species was prevented not only by ever-present deer and cattle but also by lack of sunlight, which was blocked out by the dense stands of pine, juniper, and other large shrubs and trees that formerly would have been thinned and opened by periodic fire. Even though cattle stocking was gradually reduced (and in the Gila Wilderness would eventually be nearly eliminated), the supply of browse would continue to decline and with it the deer that depended on it. Management should thus be concerned not with further protecting the deer by hunting restrictions, predator control, and refuges, but with recreating open conditions that would favor the reproduction of browse species and with holding deer populations in check, by predation and by more liberal hunting regulations, so they would not interfere with new reproduction. This would result in a population level lower than the peak level of the 1920s but considerably higher than it had been when Leopold first came to the Southwest in 1909 or than it would be if the herd were allowed to seek its own level by depleting its forage supply.[47]

The deer had probably been at about their lowest ebb, just beginning their upswing, when Leopold first arrived in southwestern game fields in 1909. It did not take him long to conclude that the white man was largely responsible for the dissolution of the habitat and the wholesale depletion of wildlife. It had occurred to him that climatic change might also be a factor, but since there was no clear evidence for it at the time and since most observed changes could also be explained as the result of man's activities, he focused on the role of human beings in the change. What man had destroyed unthinkingly he ought to be able to restore by intelligent, self-directive effort if he were to maintain a permanent civilization in harmony with the southwestern environment. Emphasizing human responsibility and management techniques as he did, it is hardly surprising that Leopold should have regarded the restoration of deer on the Gila as testimony to the effectiveness of law enforcement, predator control, and refuges.

That these manipulations should have appeared more significant than the availability of browse in restoring the herd is owing not only to Leopold's emphasis on responsibility and techniques but also to his focus on "factors" rather than on "influences," and in particular to his conception of limiting factors. Thus, he was not disposed to rate the availability of browse, or of water, cover, salt, etc.—the welfare factors, which were adequate largely as a result of forest protection and livestock developments —as positive inducements to productivity. Rather, he would simply disregard them as possible limiting factors (at least until "Pineries and Deer") and focus instead on controlling the limiting effects of illegal hunting, predators, and barren does. Hence law enforcement, predator

47. This interpretation is based on Leopold's writings in the 1940s; subsequent writings of his son A. Starker Leopold; Stewart, "Historical Background of the Black Range"; and Wendell G. Swank, *The Mule Deer in Arizona Chaparral*, Arizona Game and Fish Dept. (Wildlife Bull. No. 3, Feb. 1958).

control, and refuges would appear most responsible for the phenomenal rise in the deer population.

It may be instructive to glance at Leopold's diagram of factors of productivity in mule deer (Figure 1) and note that he not only pictured the welfare factors as negative values (poor food, poor water, poor coverts) but also for some unexplained reason lumped them as a single short line on the drawing, whereas he gave the decimating factors—hunting, predators, etc.—separate and longer lines each. While not entirely omitting mention of environmental influences such as cultivation, fire, and grazing, he tucked them off in a corner of the diagram where they could not directly affect his population curves. It was a drawing that, whether consciously or not, emphasized the relative ease of restoring deer populations by rather simple manipulations of the decimating factors. By the late 1930s Leopold would move away from this emphasis on decimating factors, though it would require decades before wildlife ecologists would be able to state with assurance that population growth is far more dependent on food, water, cover, and other habitat conditions (which are dramatically altered by Leopold's "environmental influences") than on factors such as hunting and predation that cause mortality.[48]

The productivity diagram and other diagrams and data from the southwestern manuscripts were incorporated into Leopold's new book, *Game Management*. Among the diagrams was one showing the sensitivity of the population curve for deer under various types of management, including the theoretical increase that was possible given the breeding potential of deer, if unchecked by the factors of productivity. Events on the Gila and the Kaibab had in-

48. See for example Aaron N. Moen, *Wildlife Ecology* (San Francisco, 1973), pp. 404–5: "Consideration of the productivity of a living animal rather than factors that cause mortality is much more advantageous in a population analysis because of the wide variation in potential productivity. . . . *The population growth in each generation is far more dependent on conditions that affect the productivity of the living animals than on the number of animals that are removed by harvest or natural causes.*"

dicated how nearly actual the theoretical increase could be. Yet in *Game Management* Leopold viewed the Gila and the Kaibab not as cause for alarm but as illustrations of the effectiveness of management. He did not discuss environmental influences or the extent to which the processes of vegetative change in the Southwest had been responsible for the fortunes of the deer population, nor did he treat overpopulation of deer and damage to vegetation as a new problem in game management, one that would require special attention. He wrote an entire chapter on refuges, for example, citing New Mexico as a state where the actual pattern of refuges approached the theoretical ideal, without once acknowledging that deer might overpopulate a refuge or that in the Gila National Forest alone at least four refuges had had to be abandoned because of excess deer. Similarly, in his chapters on control of hunting and of predators, he did not mention the problem of excess deer.

Leopold achieved part of the remarkable clarity of *Game Management* by simply omitting two of the most troublesome problems in "Southwestern Game Fields": a consideration of environmental influences, such as he had attempted in his chapter "The Virgin Southwest and What the White Man Has Done to It," and an attempt to determine criteria of proper stocking and productivity with reference to a particular species on a particular range. Thus he abandoned precisely those aspects of his earlier study that could have contributed to a better understanding of the deer–environment equation. This anomaly is strikingly evident in his continued use of the foresters' concept of normal productivity.

By the early 1920s, as we have seen, Leopold was already arguing against the Forest Service policy of normal stocking with respect to cattle, because the policy meant tolerating a certain amount of overutilization in limited areas and of the most palatable plants in order to secure satisfactory utilization of the range as a whole. In rough topography even localized overutilization was apt to cause erosion. Concerned primarily with maintaining the quality of the range, he had favored using the virgin condition of the watershed, rather than the capacity of the available forage to support cattle, as a criterion of proper stocking. Yet in his manuscripts on deer management in the Southwest, written at a time when he apparently did not conceive of deer overbrowsing a range or causing erosion, he had defined proper stocking as normal stocking or the most deer a range could be expected to produce.

Had he been in Black Canyon and seen the damage, and had he been convinced that deer as well as cattle had caused it, he must surely have abandoned the concept of normal productivity. Yet, although he now admitted that the Gila was not "truly normal" and hence not an adequate yardstick for measuring proper stocking and productivity, he continued to use data from the Gila to illustrate the concept, either *assuming* a stable herd (although he knew it had not been stable), or using pre-1927 figures, or noting "an as yet unanalyzed disturbance of productivity," as reflected in unbalanced sex and age ratios. He still defined normal stocking and productivity in terms of herd composition and kill ratios, rather than in terms of range quality. His focus, in other words, was on the species more than on the environment. The lessons of the Gila had not yet burned into his consciousness.[49]

In *Game Management* Leopold employed an image representative of the state of his thinking at the time—the image of a game of chess:

> The game manager who observes, appraises, and manipulates these half-known properties . . . of wild creatures, is playing a game of chess with nature. He but dimly sees the board, the men, or the rules. He can be sure of only two things: for intricacy and interest, any other game pales into insignificance; he must win if wild life is to be restored [p. 123].

49. See *Game Management* (New York, 1933), Ch. VIII, esp. p. 179.

4

Means and Ends: The 1930s

In his attitude toward the conservation of deer in the Southwest Aldo Leopold passed through a personal, telescoped version of the five-stage historical sequence of 'controls' that he was to outline in *Game Management*, a sequence beginning with restriction of hunting and culminating in environmental controls. Environmental control as he saw it was a matter of selectively manipulating the factors affecting productivity—hunting, predators, food, water, cover—in a scientific effort to restore populations of wild game and to produce a sustained surplus for hunting. He had sought to illustrate this approach in "Southwestern Game Fields" with reference to deer on the Gila. One by one, events on the Gila had called into question virtually all of his original tenets about conservation of deer. Far from discouraging him, however, the Gila experience strengthened his conviction of the potential of environmental control, even as it led him to stress the necessity of research-based experimentation on different types of range. During the 1930s Leopold promoted research and experimentation on techniques of environmental control for deer in Wisconsin, but he began shifting his attention more to the elusive "influences" of the environment and rethinking the objectives of environmental management.

Wisconsin Deer and Deer Policy

The status of the deer population in Wisconsin when Leopold arrived in 1924 was relatively the same as it had been in the Southwest when he had arrived in 1909. It was generally believed that deer were on the decline, and fear was expressed for their survival. "Deer are destined sooner or later to cease to be a game animal in Wisconsin," said the biennial report of the conservation department for 1921–1922. "Civilization is crowding them farther and farther back in narrower quarters and hunters are increasing in number each year, all of which casts a gloomy horoscope for the future of the once-abundant game animals in Wisconsin." In fact, deer populations were increasing in the northern part of the state, just as they had been in the brushfields of southern Arizona and New Mexico around 1909, but it is doubtful if Leopold or anyone had any clearer notion of what was happening in Wisconsin than he and others had had in the Southwest.

There was public clamor, in Wisconsin as in New Mexico, for strict law enforcement, closed seasons, predator bounties, and deer refuges. Starting in 1925 Wisconsin actually adopted a closed season on deer in every odd-numbered year. But Leopold did not become a leader in a public campaign for deer protection, as he had in New Mexico. Instead he became involved in the effort to establish a nonpolitical conservation commission for Wisconsin in 1927; then, as we shall see, he sought at every opportunity to encourage the commissioners to move in the direction of environmental management and research on native game, especially deer. That Leopold addressed himself to management and research rather than to game protection is owing in part to his own stage of intellectual development and also perhaps to a conviction, born of his experiences in the Southwest, that public officials needed advance preparation to lead the public to the next stage.

Leopold first seriously examined the deer situation in Wisconsin when he surveyed the game ranges of the north-central states during 1928–1930. Although he spent only a few days of his travels in deer range and did not attempt a thorough appraisal of the sort he had begun in the Southwest, he did come to a rudimentary understanding of trends in deer productivity in relation to changes in land use.

From the accounts of early explorers and naturalists he learned that, although white-tailed deer had once been indigenous to the entire north-central region, except perhaps the northern shore of Lake Superior, they were relatively less abundant in the vast forests of the north than in the central part of the region (southern Wisconsin, northern Illinois, etc.), an area characterized by belts of hardwood timber and brush interspersed with prairies and oak savannas, and having milder winters. It was the latter region of fertile soils and favorable interspersion of types that Leopold regarded as the qualitative center of the original deer range. But as this rich central area was settled and cleared for agriculture in the mid-nineteenth century, the deer were "crowded back into the poorer margins of the region." The poor, sandy soils of the north, he explained, were not capable of supporting as dense a population of deer, yet the number of deer did increase somewhat in the north after about 1870. In his report on the game survey Leopold did not associate the increase in the north with the extensive logging of white pine during 1870–1910, although years later he would explain that the deer had "followed the slashings," thriving on downed tops and on new brush that sprouted after forests were cut. The deer population had been limited, however, by market hunters supplying the logging camps and, he thought, by numerous predators and by the destructive effects of repeated, large-scale slash fires in the cutovers. Better law enforcement, predator bounties, and fire protection—controls instituted by the state after about 1915—thus seemed, as of about 1930, to have been responsible for the recent observed increase in the north and also for a new southward encroachment of deer into the sand counties of central Wisconsin and the bottomlands along the lower Wisconsin River, in areas where agriculture had failed and farms were being abandoned. As in the Southwest, man's conscious efforts at control of limiting factors weighed more heavily than the fortuitous availability of browse in Leopold's interpretation.[1]

The greatest future threat to deer abundance in Wisconsin, as Leopold saw it around 1930, was the continued logging of lowland white cedar swamps, which were important to the herd as yarding areas—sheltered places in which deer congregated during periods of deep snow and severe winter weather. But there were too many things that

1. "Game Survey of Wisconsin," typescript, 1 Oct. 1929, 167 pp., LP 6B11; *Report on a Game Survey of the North Central States* (Madison, 1931), pp. 193–99.

were not known about deer yards. How important were white cedar swamps as compared with spruce swamps? How much cutting could cedar withstand and still reproduce? How many yards remained in Wisconsin, and from what radius would deer travel to reach a yard? How far south were yards needed? Was it possible that coniferous swamps were not required for yarding, if food and shelter were otherwise available? Could upland yarding areas be created by proper interspersion of coniferous plantations and hardwood or brush feed? Leopold was quite sure that solid coniferous plantations, of the type already being cultivated in Michigan, would ultimately exclude deer, but just what constituted proper dispersion of plantings? These questions were inspired by research already begun in Michigan and other states, but they needed to be answered to apply to the conditions on the deer range in Wisconsin. The state would be embarking on a large-scale planting program, he noted, and the forestry authorities might be willing to modify their plans somewhat in the interests of deer, if only research were to disclose how to go about it. The urgent need was thus for leadership at the state level in research and demonstration.

Leopold's outlook on questions about deer and forests was peculiarly a product of his experiences in the Southwest, his efforts to forge the comprehensive, forward-looking game policy the Seventeenth American Game Conference adopted in December 1930, and his consciousness of certain contemporary economic realities in the north woods. Among other unpublished manuscripts found in his desk when he died was a handwritten fragment titled "The Game Policy in Operation," which was written about 1930 and contained a "mind's eye description" of what conditions on the deer range in the north woods might be some two decades hence if the American Game Policy were carried out:

> Let us begin with the deer range in the North Woods. The area will be divided into large units or blocks, each owned and operated for combined forestry and game management by the state or county, by a licensed lumber

company or by a licensed club. You will be able to hunt on
a state or county forest for a small fee, but you will have
to draw a ticket which tells you which one. The toll on a
Lumber holding will be higher, and the membership in a
Club very much higher. All will have more deer than now.
The open season will be months instead of days. . . .
Whenever the year's allotted kill on any one forest has been
made, no more tickets will be given out for that unit until
the next year.

Buck laws will prevail as at present. Periodically, however, the old does will be culled out by the administrative
force through the use of "mercy bullets," which permit
revival in case an error has been made. . . .

Each unit, public or private, will be under the super-
vision of a resident manager technically trained in both
game and forestry. Their individual success in various phases
of each will be widely known to the sporting public, just as
the batting averages of professional athletes are now
[known].

Toward the south edge where the deer country is thin,
and in other localities where stray bullets are especially
dangerous, hunting will be confined to bow-and-arrow, or
possibly to other short-range projectiles yet to be revived
or invented.

The whole deer country will be spattered with refuges both
public and private, but these will have more importance
as assured eating-places than as assured hiding-places.
Every wintering yard will be carefully kept.

Predator control will be moderate on public lands, but
private owners will tend to make it too stringent, and will
have to be controlled through their licenses. Control will
be regulated by density, more than by species, the public
policy being to maintain a certain low density of predators
in some fixed relation to a certain high density of game.

The state will, in short, regulate private enterprise in
deer management, partly by law and by license renewal,
and partly by the direct competition of public ownership.

The public problem will not be so much to conserve
deer; but to so regulate the crowd as to keep deer hunting
safe and enjoyable.

The foregoing describes conditions on the better forest soils. The poorer soils will be idle both as to game and forestry.[2]

Clearly, Aldo Leopold envisioned integration of wildlife management and forestry on both private and public lands by professional land managers. He assumed that the management of deer for hunting would be a profitable private enterprise, in part because of much longer open seasons; hence, his insistence on compensation to the landholder in the form of hunting fees or club membership. Such compensation was vigorously challenged at the American Game Conference in 1930 during debate on the policy proposed by Leopold's committee, but it was, significantly, sustained by majority vote. For Leopold, this new interest in compensation to the landholder represented quite a shift in emphasis from his days in the Southwest, when he had argued in favor of game management on the national forests, "the last free hunting grounds of the nation," in order to ensure democracy in sport for the common man.

The system of controlled hunting by management units was not unlike his proposal for the national forests as early as 1918. His aim in advocating controls was as much to provide quality hunting over a long season as to insure proper distribution of kill. He seems never to have questioned whether enough hunters possessed the skills or determination needed to bag a deer under natural conditions, when the deer were not frenzied by hundreds of other hunters. He did not envision any change in the traditional one-buck law. Mercy bullets, like his earlier suggestion that predators be allowed to trim some of the superannuated does on the overstocked Gila, were one more shift to avoid the uncomfortable thought of a season on does.[3]

2. "The Game Policy in Operation," holograph, c. 1930, 4 pp., LP 6B16.

3. The idea of using mercy bullets had apparently come from P. S. Lovejoy of the Michigan Department of Conservation, who wanted to use hypodermic bullets in order to remove deer to other parts of the state where there was more browse

Refuges would function not so much to protect deer from the hunter as in the Gila, but, Leopold seemed to imply, to provide islands of properly managed habitat in a sea of exploited acres. Although this function would have been valuable given the environmental conditions of the 1930s, such refuges would have been unnecessary with the extent of environmental management he expected by 1950. His conception of private attitudes and public policy on predator control was remarkably well balanced. The final observation regarding idle lands clearly bore the mark of the times: Tax reversion of cutover or worn-out lands was commonplace in Wisconsin around 1930. In a policy based largely on private enterprise and profitableness, such lands were of little account. As public funds became available during the 1930s for reforestation and watershed work, however, Leopold would begin advocating development of game habitat on what he sometimes called "waste land."

Thus during the decade of the thirties Leopold would argue repeatedly for research and demonstration at the state level on integration of deer management and forestry. His emphasis on the leadership role of the state in research and demonstration as well as in regulation was consistent with his efforts to strengthen state game departments in the Southwest. It was even more crucial in Wisconsin where the state conservation department had responsibility for forestry as well as game and where national forest administration was still in its infancy.

In addition to state forest and park lands, the department had at least some regulatory authority over several million acres of county forests and untold acreage of privately owned cutover lands that were eligible for state aid under Wisconsin's forest crop tax law of 1927. Taxpayers of the entire state, including citizens of the nonforested

---

available, and also to determine sex and age classes and other characteristics of deer. Leopold indicated he had been trying to persuade officials in Pennsylvania to use the bullets to cull superannuated does, as an alternative to doe seasons which were apparently resulting in the deaths of as many young bucks and fawns as old does. See AL to E. C. Dill, 19 Jan. 1931, LP 3B6.

southern counties, were funding the forest crop law and would be paying, through a forestry mill tax, for fire protection, reforestation, and administration of the millions of acres of tax-delinquent lands in central and northern Wisconsin that were destined to be governmentally owned. These citizens, Leopold believed, should have "at least one return from their investment, the right to hunt"; and indeed the legislature, at the insistence of the Izaak Walton League, had made specific provision in the forest crop law for public hunting on all lands entered under it. But that right would become meaningless unless constructive steps were taken promptly to maintain populations of deer, the principal game species of the forest.[4] Thus the state conservation department presumably had both the authority and, Leopold believed, the responsibility not only to initiate research on deer but also to implement the findings through adjustments in forestry policy. But as he was to discover with some bitterness, most public authorities in the 1930s shared neither his concerns nor his sense of priorities.

Leopold was particularly dismayed by the short shrift given to native wildlife. Outlays of the conservation department, derived from the sale of deer tags and other hunting and fishing licenses, went almost entirely for law enforcement, fish hatcheries, bounties on predatory animals, forest fire protection, and, after creation of a game division in 1928, the establishment of state game farms for the artificial propagation of exotic birds, principally ring-necked pheasants. Factfinding on native game and environmental conditions was relegated to a research bureau composed of department officials and of unpaid advisors, including Leopold and a number of university professors. The bureau initiated promising research on the prairie chicken and urged another top-priority study on requirements for deer habitat, but by 1933, when the chair of game management was created for Leopold at the Univer-

4. AL to Dr. Merritt L. Jones, 24 Oct. 1932; L. J. Cole to Dr. Merritt L. Jones, 24 Oct. 1932; and "Game Research Program" and "Sample Projects," typescript, 20 Oct. 1932, 6 pp. (probably written by AL); all in LP 4B7.

sity of Wisconsin, state funds for research were eliminated
entirely.[5]

In his new position at the university Leopold sought
from the start to initiate a cooperative deer study. The
conservation department was even then working up specifications for a permanent system of deer refuges, and Leopold, undoubtedly mindful of the lessons of Black Canyon and the Kaibab, stressed the necessity of local research
to provide a sounder basis for such decisions. In fact, he attempted to find an arrangement whereby Sigurd Olson, a
biologist intimately acquainted with the boundary waters
country of northern Minnesota, could work for an advanced degree at the university while doing a study of
Wisconsin deer. The game committee of the conservation
department, of which Leopold was a member, unanimously recommended such an appointment; however, there
was no action taken by the commission, and Olson was
unable to enter the university. Nevertheless, without any
special research on habitat requirements of deer in Wisconsin, the department proceeded to establish approximately a million acres of refuges and closed areas for

5. In fiscal year 1932, for example, the department spent
more than $166,000 for the warden division, about $38,000 for
bounties and predator control (down from $77,000 the year
before), over $156,000 for fish hatcheries, and $55,000 for artificial propagation at the game farm, but only $1,260 for winter
feeding (environmental management?) and $2,681 for the re-
search bureau. Forestry, now funded from general appropriations, had a budget of more than $600,000, mostly for fire pro-
tection and suppression. See Conservation Commission, *Biennial Report*, 1931–1932, financial statement, pp. 104–6.

Leopold was particularly anxious that the conservation department take the lead in deer research not only because of its
responsibility for game and forest management and its numerous field personnel but also because the deer were in the north
woods, two hundred miles from Madison, and his university
travel funds during the depression were severely limited. Moreover, the role of private landowners, his particular concern in
the early '30s, did not seem as crucial with respect to deer as
with other wildlife such as upland game birds. "If there are
any lands in the state," he wrote of the northern forests, "that
can keep on producing game without the help of the landowner, and with only such management as the state can give,
it is these lands." ("Game Survey of Wisconsin," p. 148.)

white-tailed deer, most of which within a few decades would prove to have been, on balance, worse than useless.[6]

Although most of his recommendations concerning game policy received scant attention from the conservation commission and department during the 1930s, Leopold never ceased urging the state to begin research on and management of native game and game habitat. At the federal level, while serving as a member of the President's Committee on Wildlife Restoration during January–February 1934, he argued, as he had during the Kaibab controversy, for delegation of more responsibilities to the states, as a goad to better state–federal cooperation. The Resettlement Administration and other New Deal agencies were spending millions of dollars to purchase submarginal lands unsuited for profitable agriculture, and many conservationists and state game departments, he observed, were so eager to have some of these lands devoted to the production of wildlife that they were quite willing to acquiesce in federal ownership and control. His warning was prophetic:

> The present willingness of the states to join the stampede for federalization of game does not mean they want it. They merely have their eye on the purse now open, and they know enough not to rock the boat while the emergency money is being passed out. Their complaints will be heard later.

Leopold thought the states ought to be given administrative responsibilities on most federally purchased game lands and that they ought to assume much of the research function, with the federal agency providing financial aid and acting "as a clearing house and coordinator rather than a doer." But as the chairman, Thomas Beck, pointed out, there was little disposition in Washington to encourage state administration at that time, and the committee's correspondence indicated that no small number of state game departments was "utterly unfitted" to carry on any

6. See AL to W. F. Grimmer, "Specifications for Deer Refuges," 7 Dec. 1933, and Memorandum, W. F. Grimmer to Ralph M. Immell, 13 Dec. 1933, LP 2B10; and conversation with Sigurd Olson.

program. Although Leopold did not agree, he yielded in order to approve the committee's official report. More important than the report, he decided, was the appointment of a qualified, broad-gauged administrator for the restoration program, a man who would view the job as an opportunity to shape a new federal wildlife policy in cooperation with the states and not just as a heaven-sent chance to buy some game lands.[7] His hope came as near realization as he could possibly have dreamed when President Roosevelt appointed his associate on the committee, Jay N. Darling, as chief of the Biological Survey. Functioning as liaison between Darling and authorities in Wisconsin, Leopold became instrumental in fashioning a forward-looking, cooperative plan for a state-administered conservation district in an area that had captured his imagination years earlier, the sand country of central Wisconsin.

The peat marshes and sandy scrub oak and pine barrens of central Wisconsin were a legacy of Glacial Lake Wisconsin in the Ice Age. As recently as 1870 a nesting area for the world's largest concentration (an estimated 360 million) of the now extinct passenger pigeon, the area fell victim to the drainage dredge during an abortive agricultural flurry around the turn of the century. Farming failed, and the lowered water table left dry peat to be consumed by virtually inextinguishable fires during the 1920s. In the immediate aftermath of fire, a brief weedy interlude, by chance coinciding with the high of the grouse cycle, gave central Wisconsin the largest stand of prairie chickens anywhere in the Midwest. Also, as Leopold had observed during the game survey, deer were beginning to recapture their former ranges in the central sand area. But by 1934 the ashes that had supported such an exuberant crop of food-bearing weeds had leached out or blown away, and all that seemed capable of growing, he sadly noted, were "uncountable millions" of aspen seedlings.

7. AL to Jay N. Darling, 29 Jan. 1934, and Thomas H. Beck to AL, 16 Feb. 1934, in working files, Bureau of Sport Fisheries and Wildlife, U.S. Department of the Interior, Washington, D.C. See also Thomas H. Beck, Jay N. Darling, and Aldo Leopold, *Report of the President's Committee on Wildlife Restoration* (Washington, D. C., 1934).

Prairie chickens and even brush-loving sharptail grouse were soon excluded by the aspen, and he feared that not even deer could thrive. (In those days aspen was regarded as a "weed species," of little commercial value and low on the palatability list for deer.) The aspen thickets, in Leopold's mind, threatened to become a wildlife desert.

By 1934, it was conceded in some quarters that these lands in their original flooded condition had had more potential for wildlife, recreation, and compatible industries such as cranberry growing, fur farming, and moss gathering, than they had in their drained state for either agriculture or forestry. Most of the drained and burned-out lands had reverted to the counties for nonpayment of taxes, but the counties had no field force to manage the lands in any way. Federal Resettlement Administration funds were available for buying out and resettling submarginal farmers, and CCC and FERA (later WPA) work programs offered manpower for damming the old drainage ditches and for installing food patches, cover plantings, and other improvements for game. Leopold, with his skepticism toward New Deal alphabetical relief programs, advocated administration by the state. Under a cooperative arrangement that he helped work out in 1934 for a 100,-000-acre Central Wisconsin Conservation District, county and federal holdings were to be pooled for management by the state conservation department, the counties to share in eventual revenues. This setup appealed to him as an institutional arrangement for securing game management on tax-reverted lands that could not be used profitably by private landholders engaged in agriculture or forestry. But he was under no illusions as to the complexity of ecological problems involved in restoring such an area to productive wildlife habitat. Not least was the vexing question of what could be done on the burned area now lost to aspen.[8]

8. See "The Wisconsin River Marshes," *National Waltonian*, 2:3 (Sept. 1934), 4–5, 11; and Walter E. Scott, "The 'Old North' Returns," *Wisconsin Conservation Bulletin*, 12:4 (April 1947), 13–27. For a fuller discussion of the environmental history of the sand counties and Leopold's own relationship to the area, together with a photographic interpretation, see Charles Steinhacker and Susan Flader, *The Sand Country of Aldo Leopold* (San Francisco, 1973).

In April 1935, as the conservation department was about to prepare the actual game management plan for the Central Wisconsin Conservation District, Leopold called for an explicit statement of policy from the commission on management of the area. He was concerned that the usual reforestation practices and standard CCC operations would prevail on this area, as they prevailed over so much of the state, to the detriment of carrying capacity for game. From the start he had urged the designation *conservation district* rather than *game area*, even though the area was being acquired primarily for game, because of his belief that exclusive dedications ought to be kept to a minimum. Nevertheless he maintained that the interests of game ought to have priority in case of conflict.

Leopold cautioned against the CCC's so-called "timber stand improvement" practices, such as excessive removal of scrub oak, clearing of underbrush, and burning of brush-piles. He strongly urged against large-scale plantings of jackpine, which he understood were contemplated for the area, and suggested that plantings be limited to only the very best sites, that absolutely solid blocks be avoided in all cases, and that Norway and white pine be given prefer-ence over jackpine. All conifer plantings should be kept away from marsh edges, where game food and cover plants would grow naturally, but he pointed out that the creation of conifer coverts in the centers of the larger sand "islands" in the marshes would dovetail beautifully with game man-agement operations. On esthetic grounds, he protested against the gridironing of the area with fire lanes. Al-though he was aware of fiscal elements in the situation, it seemed to him that most CCC field officers simply pre-ferred building roads to doing technical work such as dam-ming the old drainage ditches and making plantings for game. The game technicians on the ground (some of whom were his own students) were entitled to a hearing on these matters, Leopold asserted, and he was not sure they were getting it.[9]

The administrative problems on the Central Wisconsin Conservation District were, as Leopold sensed, both insti-

9. AL to Ralph M. Immell and H. W. MacKenzie, 4 April 1935, LP 2B10.

tutional and ecological. Institutionally, in addition to conflicts within the conservation department between foresters and game men and between technicians in the field and officials in the Madison office, there is some question as to how much control the state ultimately had over work done on the area during the 1930s. Although the state was theoretically entrusted with management responsibilities, various federal agencies were financing the operation, and federal technicians swarmed over the land, each doing his own thing. According to the entente so painstakingly engineered by Leopold, federal funds were not to be used for buying out county lands except in waterfowl areas and a few other special cases. Yet by 1939 the federal government had actually gained title to most of the lands in about a 150,000-acre area. In that year President Roosevelt signed an executive order designating 41,200 acres as a federal migratory waterfowl refuge, now the Necedah National Wildlife Refuge. Officials of the state conservation department and residents of Juneau County, where the refuge was located, were up in arms over what they considered devious and illegal acquisition by the federal government of lands that were to have been administered by the state, and in particular over the federal takeover of extensive upland game acreage under the pretense of establishing a refuge for migratory waterfowl. This area of fire-sown aspen, which Leopold had envisioned as a wildlife desert, would prove to be an extraordinarily productive upland game range, especially for white-tailed deer. Estimated at six deer per square mile in 1936, the protected population on the refuge would increase to a peak of eighty deer per square mile a decade later. Yet the increase in productivity for deer was largely fortuitous. As of 1936 the value of aspen as a deer food was still not appreciated, nor would it be for at least another decade, until the initiation of systematic research on food habits of deer. Research was even harder to initiate in the 1930s than new institutional arrangements.[10]

10. Leopold's role as intermediary in setting up the district is suggested in AL to Jay N. Darling, 8 Feb. 1935, G. S. Wehrwein Papers, Box 8, State Historical Society of Wisconsin (hereafter cited as SHSW). Reaction in 1939 is reported in Russell

At the same time as he was helping to set up a cooperative arrangement for securing wildlife restoration in central Wisconsin, Leopold was formulating an outline plan for statewide game management, to be published in the first annual report (December 1934) of the Wisconsin Regional Planning Committee, under the aegis of the National Resources Board. Again he stressed the leadership responsibility of the state in fostering research, establishing game management demonstrations on private as well as public lands, and encouraging extension of game management to all suitable range through creative use of the state's regulatory powers. Public purchase of game lands and use of relief labor for land development, the hallmarks of the New Deal, he no more than mentioned. These, he observed, were controlled by forces outside the scope of his report and could be brought to an end as quickly as they were begun. His concern was with building permanent foundations.[11]

B. Pyre, "State Will Fight Federal 'Grab' of Game Lands," *The Wisconsin State Journal* (Madison), 11 April 1939; and "Wisconsin Balks at Signing Treaty with Uncle Sam for Huge Land Tract," *The Milwaukee Journal*, 18 June 1939. Population estimates are from F. R. Martin and L. W. Krefting, "The Necedah Refuge Deer Irruption," *Journal of Wildlife Management*, 17:2 (April 1953), 166–76. No evidence has been found concerning Leopold's role or reaction in 1939.

The federal government did offer the remainder of its holdings (more than 100,000 acres west and north of the refuge) for administration by the state under a "treaty" arrangement, but the conservation department bitterly rejected the terms as "impossible," and it was not until July 1940 that a satisfactory lease agreement was finally reached for state administration of the Central Wisconsin Conservation Area (CWCA). As it happened, federal retention of control over the Necedah portion of the area enabled the state to conduct some highly significant demonstrations in controlled deer hunting during 1946 and 1947, in cooperation with the federal government, at a time when the conservation commission was prohibited by law from limiting the issuance of hunting permits on other lands in the state.

11. "An Outline Plan for Game Management in Wisconsin," in *A Study of Wisconsin: Its Resources, its Physical, Social and Economic Background* (First Annual Report, Wisconsin Regional Planning Committee, Dec. 1934), pp. 243–55.

The most important and the most sadly neglected of the permanent foundations was research. The total annual game budget in Wisconsin, exclusive of indirect expenditures for relief labor or land, amounted to about $185,000 ($125,000 expended by the conservation department and the remainder by the university and various federal agencies). Of this sum, Leopold pointed out, only $4,000, or a little over 2 per cent, was devoted to research, and that entire amount represented research under his own supervision at the university. In his proposed schedule of rudimentary research needs, research on deer ranked highest.

Leopold estimated that it would require five man-years of rudimentary research on deer in Wisconsin to identify and weigh the factors inhibiting their increase, to test the techniques of control on a sample area, and to run down "side-lines." (The goal was still to increase populations.) In view of the hundreds of man-years of research that have been devoted to deer and other game species in Wisconsin since the 1940s, Leopold's estimate may seem naive. But it is a measure of his optimism at the time about increasing game populations by means of relatively simple environmental controls. Even after he began to realize that populations of upland bird species were not necessarily going to respond to simple food and cover plantings but would require years of intensive ecological research, he still continued to refer to deer, by contrast, as an "easy" species to manage. Deer populations did not seem subject to the mysterious vicissitudes of cyclic phenomena or other enigmas, which conspired to thwart the best-laid plans of the manager of game birds; they could be increased with relative ease. Two decades of experience had taught him that. His vigorous plea for funding of research on techniques of environmental control for deer was thus fired by his confidence that quick dividends would follow.

Shortly after Jay Darling took over the Biological Survey in 1934, Leopold began working with him to set up the Cooperative Wildlife Research Unit Program in nine land-grant colleges. Leopold submitted a proposal for Wisconsin to participate in the program; he recommended that the bulk of the federal contribution would go for a biologist from the Biological Survey to begin a deer study, while the conservation commission would put its money into projects it had wanted to initiate anyway, such as a farmers' handbook and a warden training camp. He won strong support from the university administration, but the commission declined to participate. The commissioners resolved "to advise Mr. Aldo Leopold . . . to the effect that research is considered distinctly a function of the University of Wisconsin and the United States Biological Survey." Wisconsin lost its chance to get one of the nine units, and never did get one until 1972.[12]

We can infer from the respectfully critical tone of Leopold's communications to the conservation commission during the 1930s and from responses such as the above, as well as from the testimony of individuals who were on the scene during those years, that there existed definite tensions between the university and the state conservation department, and also a degree of personal antagonism between Leopold and H. W. MacKenzie, director of the department from 1933 to 1942. A warden of the old school, MacKenzie neither understood nor appreciated Leopold's emphasis on scientific research and habitat management, and it is said he hated Leopold's guts.[13] Compared with other states in the nation, Wisconsin's conservation department was in many ways progressive, especially in forestry matters, but neighboring states such as Michigan,

12. H. W. MacKenzie to AL, 9 May 1935, stapled with AL's "Report to Wisconsin Conservation Commission on Cooperative Research Program," typescript, n.d., 4 pp., and Chris L. Christensen to Jay N. Darling, 23 May 1935, LP 2B8. The commission was short of funds, it is true, but the reason for the shortage, ironically, was that the state would probably be unable to have an open season on deer, partridge, or ducks in 1935, and hence no income from licenses. Leopold, with his faith in the potential of research for disclosing techniques to produce more shootable game, understandably might have felt that the present scarcity of game made immediate funding of research all the more imperative and economically justifiable.

13. Discussions with Noble Clark, Joseph Hickey, Robert McCabe, Arthur Hasler, and Walter Scott. See also AL to Dean C. L. Christensen, 19 July 1935, and Memorandum for Dean Christensen, 29 July 1935, in Wildlife Ecology—General Files, Departmental, Series 9/25/3 Box 1, U.W. Archives.

Minnesota, Iowa, and Missouri were further advanced in many aspects of wildlife research and management. Leopold's cordial and cooperative relations with conservation administrators in those states and with various federal officials underscored the lack of mutual understanding in his own state. By nature he would have been unusually sensitive to such tensions as existed, yet he never gave up trying to interest the Wisconsin commission in research on native game and integration of game management with forestry.[14] Nor did he cease his efforts at the federal level to encourage programs that would give the states more responsibility, even though the principal objection by federal officials was just such imperviousness as that displayed by Wisconsin.

Deer and Dauerwald

At mid-decade, during autumn 1935, Leopold traveled in Germany on a Carl Schurz fellowship, studying German methods in forestry and wildlife management. He concentrated his attention on the history, ecology, and policy of deer–forest interrelationships and published his findings in a two-part article, "Deer and Dauerwald in Germany." Not since his studies on the life history of deer in the Southwest had he focused so closely on problems of managing deer; never had he had such an opportunity to observe the interrelated workings of deer management and forestry. It was an experience that was to strengthen his conviction of the need for deep-digging ecological re-

14. In 1937, for example, he tried again to obtain financing for a deer study, to be conducted by Richard Gerstell, who had worked on deer problems for the state game department in Pennsylvania and who now wished to study under Leopold as a graduate student at the University of Wisconsin. As Leopold explained it in a letter of 16 August 1937 to H. L. Russell, dean of the College of Agriculture: "His coming here has peculiar strategic advantages. He already carries the flavor of a *state Conservation Department* man, and would operate experimental deer pens at Poynette, the *state's* experiment station. This might help break down the present barriers to cooperation between the University game work and the Conservation Department." But once again he failed.

search; but more important, it was profoundly to influence his thinking about the objectives of wildlife management.

From a friend of southwestern days, Ward Shepard, who had studied game management in Germany a year previous, Leopold had heard about the critical deer food situation in the German forests. But he was nowhere near prepared for the extent and severity of damage deer had caused in forests, which confronted him everywhere he traveled. In the American Southwest, it had been possible to attribute range damage to overgrazing by cattle and superannuated does. In the German forests there was no livestock, and there was a policy of deliberately culling inferior deer. Leopold was brought hard against the implications of deer for forestry, and of forestry for deer:

> The observer is soon forced to the conclusion that better silviculture is possible only with a radical reform in game management. Later, as he learns to decipher what silviculture has done to the deer range, he also grasps the converse conclusion that better game management is possible only with a radical change in silviculture.

The Germans were practicing both forestry and deer management and had been for centuries; yet the two had proved mutually contradictory.[15]

The Urwald that had seen the march of Caesar's legions, Leopold explained, was predominantly mixed hardwood, with forests at higher altitudes or on poor sands tending toward conifers, and with many openings. This was ideal range and had supported an abundance and diversity of game and predators. Because of the spread of settlement and intense hunting during feudal times there was a downward trend in the deer population, but the trend was reversed dramatically after about 1400 when game management was inaugurated in the form of re-

15. "Deer and Dauerwald in Germany: I. History; II. Ecology and Policy," *Journal of Forestry,* 34:4 and 34:5 (April and May 1936), 366–75, 460–66. The historical account that follows is drawn largely from this article, with corroboration and a few added details from other of Leopold's published articles, the notes for several speeches, and notes he made while on the trip.

straints on kill, rigid patrol and penalties, predator control, and deliberate preservation of mast trees (oak and beech). The resulting large deer populations did not again decline until the exploitation of both game and forests during and after the Thirty Years War (1618–1648). Along with exploitation of deer came extermination of the large predators; and in the aftermath of forest exploitation came timber famine and a rash of short-rotation cropping expedients.

When it was discovered, around 1810, that pure spruce outyielded a mixed forest and returned a higher rent to land, the Germans stampeded to convert the forests of the entire nation to solid spruce by clearcutting and planting. The spruce mania in Germany was excruciatingly akin, in Leopold's thinking, to the then-current American infatuation with "wood factory economics" that was replanting the northern cutovers to solid jackpine. A certain amount of conifer planting was desirable in the cutovers, he acknowledged, as a counter to the vast acreages that had reverted to pure hardwoods in the aftermath of fire, but solid jackpines, like solid spruces, would only compound the wildlife problem. Deer could not subsist in a solid conifer forest devoid of openings in which sun-loving hardwood and shrub food plants might reproduce. To sustain their deer populations, the Germans resorted to artificial feeding, and they sought to ameliorate deer damage to new conifer plantings by protecting them with fences. But there were other unforeseen difficulties. By the second and third rotations of pure spruce, especially on soil outside its natural range, timber yields progressively diminished. Spruce litter did not decay and smothered all natural undergrowth. Roots failed to penetrate the soil, and windfalls increased. Insect epidemics spread rapidly where all the trees were of one species. The very productivity of the soil was diminished by podsolization—excessive acidity of the topsoil caused by lack of hardwoods, which pump up bases from the subsoil.

Challenged by soil sickness that threatened sustained-yield forest management, forestry officials in Germany shifted in 1914 to promote *Dauerwald,* or 'permanent woods'—mixed forests that would reproduce naturally.

This meant long rotations. "The human mind seems to recoil when it first faces the time element in forestry," Leopold observed, "but today the Germans calmly execute rotations of 100 and 300 years, secure in the knowledge that they represent good economics." Along with the *Dauerwald* policy came the *Naturschutz* movement, a positive and aggressive program of wildlife restoration through which, by the time of Leopold's visit, Germany had secured populations of rare or threatened birds and mammals comparable to those in the United States, despite the fact that the density of the human population in Germany was ten times as great as that of the United States. The Germans had been quick to recognize the uses of ecology and to incorporate it into official policy.

Leopold was impressed with what could be accomplished by the state, but he was not inclined to exalt Germany over America. The Germans were still managing for both high-yield timber and high-density deer; the system was still artificialized, and he did not like it a bit. Abnormally high densities of deer were still being maintained in nearly foodless woods by artificial feeding in winter, food patches for game in summer, and predator control. Adjacent agricultural lands and new forest plantings had to be protected by fencing, and spruces, larches, firs, and some unfenced hardwoods had to be individually "bundled" with dead spruce twigs wired around their trunks to prevent bark-stripping or browsing by deer.

The consequences of artificially maintained deer populations were many. Almost intolerable browsing pressure prevented reproduction of the most palatable browse species. Leopold estimated that at least two-thirds of the plant species normally occurring in the German forests had been "run out" by deer. Archer–craftsman that he was, he noted in particular the virtual absence of yew. The Nürnberg region had once supported a thriving export trade in yew bowstaves for England. "Some German foresters," he commented, "think yew succumbed to the bowstave trade, but to those who know what a small percentage of yew trees contain any staves, this sounds unlikely." In addition to yew, deer were eliminating the raspberry, blackberry, other palatable berries, and the forest game birds that required those foods, as well as wildflowers and songbirds. The official *Dauerwald* policy of mixed forests was itself being hampered by the elimination of natural reproduction.

Not least among the consequences was damage to the deer stock. Leopold had been much impressed a year previous by Ward Shepard's report to the effect that deer in the Black Forest with a buck:doe ratio of 1:6 deteriorated in both weight and antler development, while with a more natural ratio of 1:1½ and with deliberate culling of inferior individuals, they showed a 20 per cent weight increase and improved antlers. Sex ratio had seemed the key. Now, after viewing conditions in the field, he was beginning to suspect that besides sex ratio, nutritional deficiencies resulting from unnatural foods were a factor in physiological degeneration. Deterioration could probably be reversed if the artificial diets were scientifically analyzed and improved, but it could also be reversed, Leopold ventured, by restoring the natural foods of the mixed forest.

Aldo Leopold returned to the United States profoundly shocked by the realization of what too many deer could do to a forest and, conversely, what slick-and-clean silviculture could do to a deer range. Impressed as he was with the effectiveness of the state in redefining the objectives of forestry and game management and implementing the *Dauerwald* and *Naturschutz* policies, he had been even more struck with the realization that the objectives of *Dauerwald* and *Naturschutz* were being undermined by the very techniques of environmental control used to attain them. The controls were too pervasive; there was not enough play for natural forces. Leopold observed in the Germans a nostalgia for "wildness," as distinguished from mere forests or mere game, a nostalgia quite incomprehensible to most Americans:

> We Americans, in most states at least, have not yet experienced a bearless, wolfless, eagleless, catless woods. We yearn for more deer and more pines, and we shall probably get them. But do we realize that to get them, as the

Germans have, at the expense of their wild environment
and their wild enemies, is to get very little indeed? [16]

America, with its population density only one-tenth that
of Germany's, had the possibility of preserving esthetic
values in land in a way that Germany never could. To
do so, Americans would have to think less of deer and
pines and more of the integrity of the land community.
Leopold saw more clearly in Germany than he ever had
in America the ways in which deer interact with, influence, and are influenced by all aspects of the environment.
In his conclusion to "Deer and Dauerwald in Germany,"
he pleaded with Americans for "a generous policy in
building carrying capacity, and a stingy one in building
up stock."

Chequamegon and Chihuahua: The Changing Image

While Leopold was away in Germany learning the
perils of too many deer, foresters on the newly established Chequamegon National Forest in northwestern
Wisconsin were accumulating the first objective evidence
of excess deer in his own state. Making use of the abundant
manpower in CCC camps, the Forest Service conducted
organized drives to census deer on twelve sample areas of
of the forest where overbrowsing and starvation of deer
had been noted the previous winter. As a result of browse
conditions and the alarmingly high deer population indicated by the drives, an estimated 34,000, Forest Service
game men reported in December that deer numbers on
the Chequamegon needed to be reduced immediately by
fully half. They requested permission from the Wisconsin
Conservation Commission to remove 14,000 deer, including does, by a controlled hunt that winter, counting on
other mortality factors to remove the rest. [17]

The year 1935 was a closed season in Wisconsin, as all
odd-numbered years had been since 1925, so a request to
kill 14,000 deer, equivalent to roughly half the kill over

16. "Naturschutz in Germany," *Bird-Lore*, 38:2 (March–
April 1936), 102.
17. R. E. Trippensee, "Deer Problem in the National Forests in Wisconsin," typescript, 4 Dec. 1935, 8 pp., LP 4B1.

the entire state in an average open season, was certain to arouse a storm of protest. Earlier that year, in fact, a group of irate sportsmen, upset about the 1934 open season and convinced that the conservation department was allowing the extermination of deer for a profit in license fees, organized a Save the Deer Club headquartered at Hayward, only a few miles from the Chequamegon, and began agitating to keep the season closed for five years. In this exigency, the conservation commission sent a few wardens to the Chequamegon, who reported after a cursory investigation that the deer were far less numerous than claimed by the Forest Service and the scarcity of browse was not acute. The commission thereupon denied the request for reduction and was rewarded when newspapers around the state joined in rebuking the Forest Service for ever having suggested that surplus deer be removed.[18]

Ernest Swift, warden at Hayward at the time and ultimately director of the department, recalled years later that H. W. MacKenzie and some of the law enforcement men were very resentful of the Forest Service boys, "a little over-confident and a little bit smart-alecky, . . . coming in and telling them something that was so obvious they couldn't see it." The department's attitude, he thought, rather helped fuel public antagonism toward the idea of reduction. Swift himself and a few other wardens from the north country had begun to notice evidence of overbrowsing and starvation in some of the wintering yards during the early 1930s, but, like the foresters, they had failed to persuade the officials in Madison, far re-

18. See especially Wallace Grange, "Difficulties in Censusing Deer," *Game Breeder and Sportsman*, 40:1 (Jan. 1936), 10–11. Grange, former superintendent of game for the conservation department, noted the conflicting reports of the Forest Service and the department and the subsequent editorializing in the press, and pointed out that beneath all the uproar there was still a question of fact awaiting determination—the number of deer on the Chequamegon. If the Forest Service was convinced of the accurarcy of its census, it should not have taken "no" for an answer to its request for reduction, Grange stated, while the conservation department was not justified in refusing without presenting technical information to counter the Forest Service contention.

moved from the deer country. "I am telling you," Swift reminisced, "that to come into the offices from the field with a sincere belief that something was wrong without the background of quite knowing what was wrong was quite an experience." Although he realized the Forest Service "had its head to the wind in the right direction," he commented in retrospect that the Chequamegon proposal was far too drastic for the times and suggested that it may have weakened the case of fledgling field research in the public mind. In a tidy bit of analysis he explained, "Having seen deer troubles develop in other parts of the United States, the Forest Service didn't recognize it as a new thing in Wisconsin."[19]

To Aldo Leopold, just back from Germany, the news from the Chequamegon must have been startling and more foreboding by far than similar news from the Gila in the previous decade. Little more than a year earlier, in the summer of 1934, he had visited the Chequamegon and talked deer with forest officers. The discussion at that time had concerned not eliminating excess deer but means of increasing the carrying capacity for more deer. Leopold, in fact, had expressed concern about the possibility of "a general slaughter" during the 1934 hunting season if hunters were allowed access on all the new CCC-constructed fire lanes. "A very large expansion in the refuge system might ultimately ameliorate the damage," he had suggested in a rather startling reversion to his views of two decades earlier, "but this could hardly be brought about in a few months."[20]

19. Taped interview with Ernest Swift (untranscribed, c. 1963?), Conservation History Project, SHSW; Ernest Swift, *A History of Wisconsin Deer*, WCD Pub. 323 (March 1946), p. 36. Swift's analysis reflects his sympathetic understanding of the public problem, the need to educate the public, and the perils of moving farther or faster than the public would tolerate. His mild rebuke of the Forest Service, however, may also reflect his consciousness of the state–federal jurisdictional issue involved in the Chequamegon request, since he was a long-time national leader of the fight to retain jurisdiction at the state level. See Footnote 23 and James Alan Schwartz, "Ernest Swift's Use of Public Relations to Promote Conservation," M.S. research paper, University of Wisconsin, 1969.

20. AL to R. E. Trippensee, 2 Aug. 1934, LP 2B8. Although

Previous to his German experience and the report from the Chequamegon in the fall of 1935, Leopold seems never to have warned about the possibility of excess deer in Wisconsin, or indeed to have considered overutilization by deer as more than a localized phenomenon elsewhere in the nation. He was acutely conscious of the problem in Black Canyon, which had forced a road through the Gila wilderness, and he had followed closely developments on the Kaibab. He had also followed the struggles of his colleagues Seth Gordon in Pennsylvania and P. S. Lovejoy in Michigan to win acceptance in the political arena for doe seasons there. When yet another problem area developed on the Pisgah National Forest of North Carolina around 1932, Leopold had considered offering his services as a consultant, on the theory that game men in many forest areas were handicapped for lack of contact with the experiences of other areas.[21] Yet in his book

---

fire lanes represented a new hazard to deer, they also held potential for improving food conditions, thus raising carrying capacity of the range. Leopold thought white clover would be almost as good a firebreak as mineral soil; he also suggested buckwheat and other grains and warned against replanting important openings to pine. He favored trying to reestablish white cedar on suitable sites where it had been cleaned out and suggested that proper timing of cuttings to maintain a supply of felled trees in midwinter would "probably save a good many deer"—an indication, perhaps, that Leopold and forest officers were aware of stress conditions on the Chequamegon even if they were not yet talking about excess deer. A subsequent report by Trippensee on game management plans for the Manistee Purchase Unit of the Chequamegon embodied most of Leopold's suggestions.

21. AL to Ira T. Yarnall, 21 June 1932, LP 2B8. Correspondence indicating Leopold's awareness of deer problems in various areas of the country in the early 1930s is scattered throughout the Leopold Papers, but especially in LP 3B3–10.

Although the threat of too many deer nationwide became more real during the 1930s, it materialized later in Wisconsin than, for example, in Pennsylvania and Michigan. The environmental disturbances that set the stage—principally the westward march of the loggers and subsequent slash fires—occurred later in Wisconsin, and protective measures such as fire control, buck laws, bounties, and refuges were instituted later. Thus, Pennsylvania's herd had peaked at about a million in the mid-1920s and was already in process of rapid decline; Michigan

Game Management he had discussed population mechanisms and productivity factors, hunting restrictions, predator control, and refuge systems without once thinking it
necessary to deal with the phenomenon of excess deer.
His assumption was still that deer could be managed
downward as readily as upward when the need arose,
given a few man-years of research at the local level on
habitat requirements and environmental controls.

In his contribution on game management for the Wisconsin regional plan report in 1934 Leopold had specifically raised the question, "Is ultimate over-production
probable?" and he had concluded that it was not. But what
is most significant is the context in which he had raised the
question. He was still, as in some of his earliest writings
in the Southwest, thinking of the supply of shootable
game animals in relation to hunter demand, and hunter
demand was to his mind virtually insatiable. Thus, even
if the annual yield of shootable deer were increased to
25,000, which he calculated was possible given Wisconsin's acreage of deer range and potential yield per acre,
there would be only one deer per eight hunters—obviously
no threat of surplus.[22]

In 1934 Leopold had assumed the chairmanship of a
new committee of the Society of American Foresters on
forest game policy, which included such experts on problems of excess deer as Walter P. Taylor of the Biological
Survey and C. E. Rachford of the Forest Service, both of
whom had investigated conditions on the Kaibab and in
Black Canyon, and P. S. Lovejoy of the Michigan Department of Conservation. In the committee's first report,
drafted by Leopold, there had been no mention of the
problem of overutilization of ranges by deer. It is noteworthy that the committee did not deal with the problem
directly because there was a section dealing with the ju-

probably had about 500,000 deer around 1930, but its herd was
still mushrooming; while Wisconsin's herd, although known to
be increasing, was officially estimated at a mere 25,000, which
was undoubtedly too low, as the decade opened.

22. "An Outline Plan for Game Management in Wisconsin,"
p. 249. Leopold raised the question of overproduction with respect to game in general, although he discussed primarily upland game birds and dealt with deer only peripherally.

risdictional problem on the national forests, a problem
that had come to the fore because of the reluctance of
state legislatures and conservation departments to provide for adequate reduction of deer on forest areas damaged by overutilization.[23]

23. Several months before Leopold's committee was formed,
Secretary of Agriculture H. A. Wallace had flung down the
gauntlet on jurisdiction by issuing Regulation G–20–A, which
declared that the Secretary, upon recommendation of the Forester, might establish hunting regulations for national forests
in cases where he judged state laws or policies inadequate to
regulate wildlife populations and preserve other forest values.
State conservation officials across the country rallied in defense
of their common-law jurisdiction over game against threatened
encroachment by the Forest Service. Reaction against the complex of attitudes supposedly represented by G–20–A could be
read in J. Stokley Ligon's 1934 contention that the Forest Service was the most serious menace to big game in New Mexico,
as it could also be seen behind the Wisconsin commission's pre-
emptory denial of the request to deal with the problem on the
Chequamegon.

To Aldo Leopold, the jurisdictional issue was a bugbear.
He had argued against the Boone and Crockett Club's advocacy
of Forest Service jurisdiction in the Kaibab affair back in 1923,
on the grounds that state conservation departments needed to
be strengthened rather than undermined, and this was still his
principal consideration. In the meantime he had become more
than ever convinced that game could adequately be managed
only by the man on the ground. It was the part of state game
administrations, he suggested, to encourage such management
by delegating authority for details of management to the landholder, whether private or public, under suitable state regulation. In practical terms this meant to him that on national
forest lands, where rangers were already on the ground, the
states ought to delegate authority for management to the Forest
Service under suitable cooperative arrangements, while on the
vast, scattered acreages of submarginal lands being acquired
under various New Deal relief programs, such as the Central
Wisconsin Conservation District, the federal government ought
to entrust the states with management responsibilities. Leopold's section on jurisdiction was arbitrarily deleted from the
published version of the forest game policy committee's first
report, apparently because it was felt that his acknowledgment
of state jurisdiction might prejudice the Forest Service case in
the G–20–A controversy, but it did appear in the second report
of the committee in 1936. Committee reports and correspondence are in LP 2B8.

It was in the committee's second report, written in 1936 after Leopold's trip to Germany and after the flare-up over the Forest Service request for reduction of excess deer on the Chequamegon, that Leopold finally confronted directly the problem of overstocking. "The current year," he wrote, "has seen the emergence of a biological principle, perhaps long realized but not previously asserted as a positive and generalized rule for guidance in wildlife administration. It is this: overstocking range with game birds carries no invariable penalty in loss of future carrying capacity, but overstocking range with browsing mammals does." The public was as yet oblivious to this principle; their clamor for winter feeding and predator control on an overstocked deer range, he observed, "is just as insistent, sincere, and uncritical as on an understocked quail, turkey, or pheasant range." Hence the urgent need for administrative caution in avoiding overstocking of browsing mammals and for expanding local research, especially in food habits. He cited "Deer and Dauerwald" in support of his conclusion that wildlife managers needed to transfer their emphasis from "more game" to "more carrying capacity."[24]

When Leopold wrote that "the current year has seen the emergence of a biological principle," he was admitting, in effect, that he himself had suddenly come to appreciate the broader significance of the excess deer problem, though he had been involved with the problem for years. He now realized it was more than a matter of supply and demand or of management techniques and involved much larger considerations of environmental quality—what he would come to call land health. That overstocking with deer would not only deplete forage sup-

24. "Second Report of Game Policy Committee," typescript (first draft), 15 Sept. 1936, 6 pp., LP 2B8. See also *Wildlife Crops: Finding Out How to Grow Them* (Technical Committee of the American Wildlife Institute, Aldo Leopold, Chairman: Washington, D.C., 1936), p. 10: "In the last decade, research has disclosed one fact about deer and other browsing mammals worth a hundred times the total sum ever invested in all scientific work. It is this: browse-eating game can and does ruin its own range, seed-eating game does not."

plies and thereby lessen the carrying capacity of the range for deer but also change the composition and undermine the diversity of the vegetation for years to come was a startling idea. If he had thought about it, of course, he would have realized that he had long known of instances in which deer affected vegetation, just as he had long been aware of the consequences of overgrazing by cattle and elk. But he had not allowed these considerations enough weight in determining the objectives of deer management. It would be difficult to overestimate the impact of Leopold's trip to Germany, coming when it did, on his thinking about means and ends in wildlife management.

The transformation in attitudes concerning excess deer was of course only a focal point of a broader transformation around the mid-1930s not only in Leopold's thinking but in ecological ideas generally. In 1936 when Leopold drafted the second report of the game policy committee he was acutely conscious of the prevailing confusion of thought in the entire conservation field. "An intellectual revolution seems to be in process," he wrote, "the net effect of which is to vastly expand both the importance and the difficulty of the conservation idea. During this process, it is difficult to see far ahead. At any rate, it is difficult for us." What would emerge from this revolution was the biotic idea. Leopold already grasped fragments of the idea and emphasized them in the report: a new concern for the ecological penalties of overstocking; an insistence on preservation of rare or perishable features of public lands, even if it meant restriction of public access; and, perhaps most significant, a call for deliberate management in the interest of nongame wildlife, such as the wolf and the grizzly.

One of the first to integrate new research findings in wildlife, forest, and range management with emerging ecological theory was Walter P. Taylor, a member of Leopold's committee and also president of the Ecological Society of America, whose paper, "Significance of the Biotic Community in Ecological Studies," was published in September 1935. Taylor was concerned with the way ecologists conceptualized the relationship between organism and environment. Many plant ecologists, he observed, regarded animals as biotic factors external to the plant community, while animal ecologists tended to think of plants as simply a portion of the habitat in which animal communities lived. During his earlier work, Leopold himself had tended to think of deer as influenced by various environmental factors rather than as themselves a functional part of the environment. Taylor preferred to think of the biotic community, or even "the organic-inorganic complex complete," rather than plants alone or animals alone, as the real entity, the "organism," and he regarded it as a single system of material and energy.[25] In philosophical terms Leopold as early as 1923 had conceived of the whole earth and the smallest particle thereof as a living being, an organism. But it was not until the convergence of research and theory in the mid-1930s that an ecological basis for the organic concept began to emerge and its implications for wildlife management became apparent to him. He now was beginning to think more in terms of preservation of the system as a whole, of its capacity for smooth functioning and internal self-renewal, rather than in terms of augmenting the supply of a particular resource.

Leopold's thinking may well have been influenced by his association with men like Taylor and other ecologists and land managers, but the penetration, conviction, and clarity of his mature thought owed even more to his experiences in the field. Field experiences for him were not only an opportunity for sustained observation but also for reflection, for putting together the pieces. In the field, he was often able to see new relationships among what had formerly been discordant facts. The resulting clarification extended to other aspects of his thinking as well.

25. Walter P. Taylor, "Significance of the Biotic Community in Ecological Studies," *Quarterly Review of Biology*, 10:3 (Sept. 1935), 291–307. Taylor makes no mention of Leopold in the article, probably because he did not consider Leopold one of the "ecologists." On Leopold's personal copy of the article appears his handwritten notation, "Examples of plants dependent on animals."

This process was perhaps most dramatically illustrated by his experience in Germany, but that trip was followed a year later by another field experience that helped complete the transformation in his thinking.

In September 1936 Leopold went on a pack trip along the Río Gavilán in the Chihuahua sierra of northern Mexico. It must have been as satisfying as his German experience was disturbing, for he was finally in the presence of a deer herd in balance with its environment. "It was here," he reflected years later, "that I first clearly realized that land is an organism, that all my life I had seen only sick land, whereas here was a biota still in perfect aboriginal health. The term 'unspoiled wilderness' took on a new meaning." [26]

Protected from settlement and overgrazing, first by Apache Indians and then in succession by Pancho Villa's bandits, depression, and unstable land policies, the Sierra Madre in Chihuahua still retained the virgin stability of its soils and the integrity of its flora and fauna. "It is ironical," Leopold mused in a published account of his observatons, "that Chihuahua, with a history and a terrain so strikingly similar to southern New Mexico and Arizona should present so lovely a picture of ecological health, whereas our own states, plastered as they are with National Forests, National Parks and all the other trappings of conservation, are so badly damaged that only tourists and others ecologically color-blind, can look upon them without a feeling of sadness and regret." He noted that the area burned over every few years but with no ill effects, "except that the pines are a bit farther apart than ours [southern New Mexico and Arizona], reproduction is scarcer, there is less juniper, and there is much less brush, including mountain mahogany—the cream of the browse feed." Deer were abundant, but not excessive. In nine days of hard hunting, Leopold and his brother Carl saw 187 deer, 50 of them antlered bucks. "Deer irruptions are unknown," he stated categorically. The deer thrived in the midst of their natural predators, wolves and mountain lions, in a ratio that he believed was not much different

26. "Foreword," typescript, 31 July 1947, 9 pp., LP 6B17.

from that in Coronado's day. Coyotes had not invaded the mountains as they had all over the United States, he observed, wondering whether wolves had kept them out.[27]

Leopold's trip in the Chihuahua sierra marked his first clear realization that deer and predators could coexist in relative equilibrium in an uncontrolled environment. By the mid-1920s, he had come to realize that wolves and mountain lions added to the diversity of wildlife and therefore ought not to be exterminated. A few years later, he had suggested that they might have a function in trimming sick or aged deer under properly controlled conditions. By the mid-1930s, he was asserting that they were threatened species, fully as important as bighorn sheep or Mearns's quail, and as such ought to be deliberately managed for. But it was the trip to Mexico that finally made him appreciate the function of predators in maintaining the health of the system—or indeed let him see what health was.

Even as he visited the area, however, it was threatened by new resettlement programs of the Mexican government and excessive hunting by American sportsmen. He did not know what the future would hold; "but in any event," he concluded, "the Sierra Madre offers us the chance to describe, and define, in actual ecological measurements, the lineaments and physiology of an unspoiled mountain landscape":

> What is the mechanism of a natural forest? A natural watershed? A natural deer herd? A natural turkey range? On our side of the line we have few or no natural samples left to measure. I can see here the opportunity for a great international research enterprise which will explain our own history and enlighten the joint task of profiting by its mistakes. (p. 146)

Leopold returned to the Sierra Madre the following year with his brother Carl and his son Starker, then tried to interest the eminent geographer Carl O. Sauer, a pro-

27. "Conservationist in Mexico," *American Forests*, 43:3 (March 1937), 118–20, 146. For a thorough appraisal of Mexican wildlife and its management see A. Starker Leopold, *Wildlife of Mexico: The Game Birds and Mammals* (Berkeley, 1959).

fessor of Starker's at the University of California at Berkeley who had done considerable field work in northern Mexico, in his idea of an international research effort, thinking the Carnegie Institute might be inclined to fund such a venture. The primary focus of the study would be the soil–water–streamflow relation in the northern Sierra Madre, as compared with the "modified" terrain of similar geologic formation in southern Arizona and New Mexico, in order to determine what the original equilibrium consisted of. Where a decade earlier, in his chapter "The Virgin Southwest and What the White Man Has Done to It," he had tried to determine the former condition through the historical record, he now had a natural laboratory where the question of biotic equilibrium could be explored. In the meantime, he had come to realize that not only did plants play a role in the biotic equilibrium and "determine" the animals that would thrive in a given area but the animals could also "determine" the plants. In his various writings about the Gila, as we have seen, he had not successfully integrated the changing fortunes of the deer herd with his general interpretation of environmental change. But now, as he indicated to Sauer, he was especially interested in studying the deer–wolf–coyote relationship, to see if he could discover the mechanism responsible for the peculiar stability of the Chihuahua deer and shed some light as well on the phenomenon of watershed equilibrium. Opportunity to launch the study never materialized, though Leopold would try several times in the ensuing years to revive the idea. He did not give up hope of one day studying the mechanisms of a truly healthy biota.[28]

* * *

In summer 1948, a few months after his father's death, Starker Leopold returned to the Sierra Madre with a scien-

28. See AL to Carl O. Sauer, 29 Dec. 1938, LP 3B3. There is no reply from Sauer in the file. Leopold's journal of the second trip has been published in *Round River: From the Journals of Aldo Leopold*, ed. Luna B. Leopold (New York, 1953), pp. 130–41. For a discussion of a subsequent attempt by Leopold to initiate a study in Chihuahua see pp. 177–80.

tific team from the Museum of Vertebrate Zoology at Berkeley, intending finally to initiate the long-term study of deer and predators in virgin equilibrium that he and his father had envisioned on their bow-and-arrow expedition. What he encountered were graded logging roads, herds of cattle and sheep, and the Río Gavilán flood-scoured and flowing sawdust. He understood then why his father had once written, "It is the part of wisdom never to revisit a wilderness," and he bade farewell in a moving essay, "Adiós, Gavilán."[29]

Rockford and Huron Mountain

Aldo Leopold, back in Wisconsin in the late 1930s, continued in his efforts to get the conservation commission to initiate deer research of the most rudimentary sort. His new awareness of the biotic penalties of overstocking, coupled with indications from the Chequamegon that excess deer were already a reality in the state, gave his pleas a new urgency if not a more sympathetic hearing. Though his efforts in the public arena were to have little immediate effect, he was to have an opportunity to work out his new approach to deer–forest management on two consulting projects on private lands.

The state conservation department did conduct some deer census drives in the mid-thirties, using CCC labor; and in view of evidence that there were more than thirty deer per section on the areas sampled, the state in 1937 went to open seasons on bucks every year, instead of only in the even years. In justifying consecutive open seasons, department officials explained that with the cessation of extensive lumbering operations in much of the north, winter food in the form of fresh slashings was definitely on the decrease, making deer damage to pine plantations and agricultural crops and eventual starvation a real threat unless larger numbers of deer were harvested. The

29. AL, "The Green Lagoons," *American Forests*, 51:8 (Aug. 1945), 376; A. Starker Leopold, "Adiós, Gavilán," *Pacific Discovery*, 11:1 (Jan.–Feb. 1949), 4–13. See also AL, "Song of the Gavilán," *Journal of Wildlife Management*, 4:3 (July 1940), 329–32.

emphasis was thus on shortage of food, a perfect though unintended setup for sentimental citizens who saw winter feeding with hay or grain as the obvious solution. That over-all food supply was in fact greatly augmented as a result of the new growth of brush on formerly cutover lands, and that deer numbers were in fact exploding, seems not to have been understood at the time.[30] Only through research at the state level, Leopold was convinced, would state officials or the public come to appreciate the threat of excess deer. But research in Wisconsin was still years off.

The first of Leopold's private consulting ventures involved a self-contained herd on a wooded agricultural estate along the breaks of the Rock River near Rockford, Illinois. It was of special scientific interest to Leopold because the area was typical of what he regarded as the pre-settlement center of abundance of deer in the Midwest. On the basis of his first field inspection, in October 1936, he guessed the density might be as high as 64 deer per square mile, including both woodlots and farmland, or twice the densest populations sustained on comparable wild areas. The estate did not yet show signs of overbrowsing, but Leopold pointed out that the population had been increasing rapidly in recent years and that the effects of the recent increase might not yet have been registered. He stated emphatically that, for the health of both the range and the deer, the herd should not be allowed to increase any more. He therefore recommended securing special state authorization for removal of any future surplus, since there was no season on deer in Illinois, and he advised that the two sexes be removed about equally: "I especially

30. See, for example, Ernest Swift, "Review of Deer Season," *Wisconsin Conservation Bulletin* (hereafter cited as *WCB*), 1:12 (Dec. 1936), 4–7; H. W. MacKenzie, "To the Citizens of Wisconsin Interested in the Deer Question," *WCB*, 2:9 (Sept. 1937), 3–9; Ernest Swift, "The Problem of Managing Wisconsin Deer," *WCB*, 4:2 (Feb. 1939), 8–27; and W. E. Scott, "Status of the Wisconsin Deer Herd," typescript, 1939, 7 pp., WCD Records, SHSW. Swift and Scott were among the earliest WCD officials to express concern about the deer problem and to argue for herd reduction and for research, yet in their public statements during the late 1930s they supported the department's attempt to tread a safe line between opposing forces.

warn against the prevalent assumption that a sex ratio of one buck to four or five does is desirable. A ratio of about 1:1 is much better for the ultimate welfare of the herd." (In the old days, when the aim was to raise maximum deer for hunting, he had emphasized the polygamous habit of deer, which rendered three-fourths of the bucks legitimate gun-fodder. Now he was principally concerned about the vigor of the herd, especially since the owner of the estate enjoyed viewing the deer rather than hunting them.) Following his own advice in "Deer and Dauerwald," he also recommended measures to build up the carrying capacity of the range to a safer margin above population, such as better dispersal of agricultural feeds, new plantings of sunflowers and vetch, and plantings of additional wild browse species like box elder, soft maple, and red dogwood.

Regular, careful observation on the part of the superintendent of the estate, landscape architect Paul B. Riis, would be required, Leopold stressed, in order to assure the continued health of the range and the herd. The most accurate and easily observed criterion of a healthy range was the degree of browsing on palatable woody plants. The nipping off of more than half the new growth on accessible palatable shrubbery should be regarded as overbrowsing, Leopold suggested, and any general browsing of unpalatable species would also spell danger. As an aid to Riis, Leopold prepared preliminary 'palatability lists' indicating plants that showed browsing and those that were untouched, plants eaten in summer only, those possibly already eliminated by deer, and others not yet classified. He specifically warned about the insidious substitution of unpalatable for palatable species and indicated that ironwood and hazel might be the main species to increase on the Rockford range as a result of overbrowsing. As criteria of a healthy herd Leopold suggested weight and antler development of individuals, dispersion of fawning dates (a wide dispersion indicating abnormality in sex ratio or physical condition), and pelage changes; and he provided Riis with what few standards of comparison were available from studies in other areas. He stressed repeatedly the potential scientific value of any studies made on the area and offered his continuing help on the project.[31]

On his second visit to the area a year later, Leopold was impressed by the amount of summer browsing, and reiterated more strongly a point he had made the year before in regard to criteria of a healthy range: Plant species that have been eliminated and replaced by less palatable species as a result of browsing pressure are just as important as plants that are damaged. Accordingly, he suggested experimental plantings of species no longer present, including yew, ferns, wahoo, and browsable-sized maple, to see if they disappeared. "I reiterate the basic principle," he wrote, "that the elimination of any plant species should not be permitted." To make a smaller number of deer produce greater viewing satisfaction for the owner, he suggested localizing them in summer on clover patches in old fields. In an agricultural region, that would hardly be esthetically objectionable.[32]

Leopold's new emphasis on summer browsing and its effect on the diversity of plant species was in marked contrast to the prevailing concern at the time over winter browsing and deer starvation. Hardly anyone worried about summer browsing because there was clearly no shortage of food for deer in the summertime. It would be at least another decade before Leopold would even begin to get any conservation officials to think seriously about the environmental effects of summer browsing. The general public has yet to understand.

The second consulting opportunity was more comprehensive. Leopold was asked to recommend a land program for the Huron Mountain Club, an exclusive group of nature lovers and sportsmen who owned a 15,000-acre tract

31. "Report to Mr. Howard D. Colman on Rockford Deer Area," typescript, 15 Oct. 1936, 4 pp., LP 2B7.

32. "Second Report to Mr. Howard D. Colman on Rockford Deer Area," typescript, 21 Sept. 1937, 5 pp., LP 2B7. Leopold continued to encourage and advise Paul Riis on his studies of deer food habits, at least until 1940. The Riis–Leopold correspondence and data were used by Lyle R. Pietsch in connection with his "White-Tailed Deer Populations in Illinois," Illinois Natural History Survey Biol. Note 34 (1954).

of virgin maple–hemlock forest on the south shore of Lake Superior west of Marquette, Michigan. He visited club lands twice during the summer of 1938, prepared a masterly report that the club printed in pamphlet form, and through conferences and correspondence helped lay the groundwork to effect his recommendations. The Huron Mountain property, he noted, would soon be one of the few large tracts of maple–hemlock forest remaining in a substantially undisturbed condition; it was of outstanding value for wilderness recreation, scientific study, wildlife conservation, and timber. The wilderness and timber values belonged wholly to the members as owners of the land, but the scientific and wildlife values, he pointed out, were in part public values. The club property was also deeply affected by the laws and policies of public agencies and by the actions of neighboring landowners, so club policies clearly could not be decided in a vacuum.[33]

The Huron Mountain lands in presettlement times were an area of relatively sparse deer population; but now, although carrying capacity of the mature forest was essentially as low as it had always been, the area was imperiled by too many deer. Plimsoll (browsing) lines were already evident on white cedar along lakes, yew was absent, and white cedar and hemlock were not reproducing. Given the preeminent value of the virgin maple–hemlock forest as a base datum for scientific study, any diminution in diversity of species or radical change in the structure of the community as a result of overbrowsing by deer would be intolerable. Yet Leopold realized that no matter what the club did on its own lands, it was vulnerable to sudden changes beyond its boundaries, especially because its boundary was very "ragged." A logging railroad and logging camp had just been completed within a mile or two of club lands, and nearby owners, including the Ford Motor Company, were exploiting their holdings by heavy cuts. The slashings and regrowth would sharply raise car-

33. See AL, "Report on Huron Mountain Club," typescript, 5 June 1938, 29 pp., and other reports and correspondence in LP 2B4. The report was printed as a pamphlet in 1938 and reprinted in *Report of Huron Mountain Wildlife Foundation 1955–1966* (Jan. 1967), pp. 40–57.

rying capacity for a few years and then leave a heavy deer population to inflict damage on club lands.

Leopold's response to the problem was many faceted, but his basic objective was the one he had outlined in "Deer and Dauerwald" and had applied at Rockford—preservation of a safe margin between carrying capacity and population, by raising carrying capacity and holding down population. Where at Rockford carrying capacity was to be increased by means of special plantings, at Huron Mountain the mechanism Leopold recommended was selective cutting to allow light for natural reproduction. In the center of the property, however, he urged reservation of a large natural area for scientific and esthetic purposes. This would not be cut, but in a buffer zone surrounding it light selective cutting would be practiced to create openings for browse reproduction and also for wildflowers and songbirds. The technique of selective cutting in northern hardwoods had only recently been developed under Leopold's colleague Raphael Zon at the Lakes States Forest Experiment Station. The traditional method of opening a forest like Huron Mountain was by clearcutting or slash, and it was that technique, still used by Ford and other neighboring landowners, that wreaked havoc with the deer population. Hence Leopold strongly urged the club to block out its holdings, if possible, by contracting for cutover land before it was logged and adding a selective-cutting stipulation. In the next few years, he even managed to persuade Ford and the Forest Service to coordinate with the club on selective cutting and buffering.[34]

34. A forestry program such as Leopold recommended had to be "sold" to the individuals involved. One of the key men, Daniel Hebard, chairman of the club Lands Committee, was not impressed with the idea of selective cutting, particularly since he had seen exploitative "selective" cutting on the Ford lands. Leopold got the private forestry division of the U.S. Forest Service to coordinate buffer zones in a survey they did of the Ford lands in summer 1939. The foresters then visited Hebard armed with the Ford survey and accounts of similar work for Sawyer Goodman of Wisconsin. This, plus the fact that Hebard, meanwhile, had learned of the esteem in which Leopold was held by Herbert L. Stoddard, who consulted on management of Hebard's plantation in Georgia, was enough to make him quite an enthusiast for selective cutting. The Forest

Logging would continue, however, and the deer population would rise, unless it could be held in check by hunting
and predation. Leopold urged the club not only to encourage hunting but also to pioneer in organizing a group
of landholders to work with the Michigan Conservation
Department in order to provide flexible wildlife administration. Such an enterprise would entail a year-round research effort dovetailed with the deer investigations cur-
rently being conducted by the state. He recommended
that the special focus for the Huron Mountain study be
deer–wolf–forest interrelationships. "Who will have the
courage," he queried, "to 'break the ice' by proposing a
rational policy toward the remnant of Michigan wolves?"
Any serious student of the northern forests, he asserted,
foresaw the heavy physical damage that would be inflicted
by excess deer, yet no one had succeeded in reversing
the governmental policy of subsidized extermination of
wolves. Many intellectuals were beginning to rebel against
blanket predator control, but they shunned involvement
in administrative affairs. "Natural resources, then," Leopold concluded, "are one of many public interests which
suffer from the current refusal of highly trained men to
enter 'politics'." Nor could the Forest Service or the state
conservation department take the lead, without antagoniz-
ing the "barbershop biologists" of the north country and
risking retaliation against other phases of their program.
But the agencies would probably be ready to move, Leopold suspected, if given backing by some citizens' group
other than the intellectuals. Here then was the opportunity for the Huron Mountain Club, not only to cease killing
of wolves and other predators on its own lands but also to
provide backing in the form of measured facts on deer–
wolf–forest interrelationships.[35]

Service, accordingly, went ahead with a detailed plan for the
club lands, closely based on the Leopold report. See LP 2B4 and
"Possibilities of Coordinated Forest Land Management for the
HMC," Division of State and Private Forestry, USFS, Milwaukee (copy at U.W. Dept. of Wildlife Ecology), 1940.

35. A bounty on wolves was instituted by Michigan's first
legislature in 1838. From 1935 to 1960, more than 700 timber
wolves were bountied, and by the time the species was finally
given complete protection in 1965 it had been effectively ex-

Leopold's recommendations were accepted by the club, and over the years nearly all of them were carried out. As logging encroached throughout the Upper Peninsula the virgin wilderness of the Huron Mountain Club assumed ever greater value for science, and the club's positive encouragement of research led to scores of significant studies.[36] The "Report on Huron Mountain Club," with its case for encouragement of private responsibility in developing both public and private values inherent in the land and with its comprehensive, cooperative program for meeting the threat posed by excess deer, is representative of Aldo Leopold's finest thinking on environmental management.

Transmutation of Values

On his trek from Blue River and Black Canyon to Rockford and Huron Mountain, Aldo Leopold passed not

terminated. Donald Douglass, chief of the game division of the Michigan DNR, speaking at the North American Wildlife and Natural Resources Conference in 1970, indicated that there were probably less than a dozen individuals remaining and no evidence of social groups that could reproduce. The individual wolves were widely scattered over the Upper Peninsula, but the most consistent reports, he noted, came from the Huron Mountain Club. "We are considering 'tip-toeing boldly' into a program of supporting the restoration of the wolf in the Upper Peninsula, and we would be very glad to follow a private enterprise effort to gain acceptance of the idea," he said. See *Proceedings of a Symposium on Wolf Management in Selected Areas of North America*, S. E. Jorgensen et al., eds. (Twin Cities: USDI–BSFW, 24 March 1970), pp. 6–8.

36. For the immediate response to Leopold's recommendations see "Memorandum—Meeting of the Forest and Wildlife Committee, Huron Mountain Club," typescript, 5 Sept. 1938, 8 pp. (copy courtesy of William P. Harris, Jr.). The committee was emphatic about ceasing all predator control activities. The club provided scholarship funds for a graduate student, Richard H. Manville, to make a three-year study of wildlife populations on club lands, and Leopold advised Manville during the course of his studies. See Manville, "Report on Wildlife Studies at the HMC, 1939–1942," mimeographed (copy at U.W. Dept. of Wildlife Ecology), 1 Aug. 1942, 165 pp. Other studies are summarized and a bibliography provided in the *Report of Huron Mountain Wildlife Foundation 1955–1966*.

only through a telescoped version of the historical sequence of game management techniques that he outlined in *Game Management* but also through a sequence of objectives determined somewhat by the changing nature of the deer problem itself and of his perception of it. During his days in the Southwest when the problem was depletion of deer, he directed his efforts first toward retarding depletion and then toward increasing herd numbers by controlling the various factors (hunting, predators, buck shortage or barren does, food and water, etc.) that held down the natural productivity of deer. Although some of his controls were "environmental," his focus was on productivity of the species. As deer numbers increased and local over-utilization problems developed on the Kaibab and in the Gila, Leopold began to perceive the problem of deer restoration as a matter less of deer productivity and more of range productivity, or carrying capacity. The challenge was to regulate livestock grazing more stringently, to cull the old does, and to use research to find some way to stimulate browse reproduction in order to support more deer. In Wisconsin it was a matter of maintaining wintering yards and paying attention during reforestation to proper dispersion of food and cover types.

As late as 1935, Leopold still did not perceive excess deer as more than a localized, temporary problem. He simply assumed that game managers would be able to develop appropriate techniques for maintaining populations of wild game in equilibrium with their environment. Although his objective was no longer, if indeed it ever had been, merely production of maximum game for hunting, he still measured success in those terms. "The production of a shootable surplus," he had asserted in his contribution to the Wisconsin regional plan report (1934), "is the acid test of the sufficiency of a conservation system." The implication was that songbirds, wildflowers, and other "noneconomic" members of an esthetically pleasing, diverse community would automatically thrive if environmental conditions existed that were needed to produce shootable game.

It required confrontation with managed deer in the managed forests of Germany to make Leopold appreciate fully the real threat of excess deer, and the extent to which objectives as well as techniques of management were involved in the excess deer question. And it required experience in the diverse, fully integrated biotic community of the Chihuahua sierra to make him appreciate the alternative of land health. The threat of excess deer was a threat to the diversity and stability of the environment itself, to its capacity for smooth functioning and internal self-renewal, or health, which he now regarded as essential. On his return from Germany and Chihuahua, Leopold shifted from an emphasis on environmental management in order to increase deer populations for sport hunting to an emphasis on environmental management in order to provide a safe margin between the carrying capacity of the environment and deer population, with controlled recreational hunting as a management tool for maintaining appropriate deer populations. The purpose of environmental management in the broadest sense, as distinguished from management of deer herds or deer ranges, was not so much to increase productivity of or carrying capacity for a single species as to rebuild a diverse, healthy environment.

Not many months after his return from Germany, Leopold gave a talk to the Chamberlain Science Club at Beloit College on "Means and Ends in Wildlife Management," in which he mused on the peculiarities of a profession grounded in science but with an output measured in esthetic satisfaction:

> There seem to be few fields of research where the means are so largely of the brain, but the ends so largely of the heart. In this sense the wild life manager is perforce a dual personality. Whether he achieves any degree of consistency as between his tools and his objectives, we must leave for others to judge.[37]

Three years earlier, in his book *Game Management* and in his address on "The Conservation Ethic," Leopold had exuded optimism over the sheer possibilities of environmental control, if only it be used to socially desirable ends. Now he was telling science students at Beloit that wildlife

37. "Means and Ends in Wildlife Management," holograph, 5 May 1936, 5 pp., LP 6B16.

managers had discovered they were unable to replace natural equilibria with artificial ones by means of scientific 'controls' and would not want to even if they could. Within another three years, in his "Biotic View of Land," he would buttress his preference for diverse, naturally maintained equilibria with ecological as well as with ethical and esthetic arguments.

The biotic idea had important implications for land use. Manmade changes in the biotic pyramid were of a different order from evolutionary changes, Leopold observed in his 1939 paper, and biotas, moreover, seemed to differ in their capacity to sustain violent conversions to human occupancy. Western Europe, for example, despite vast changes in the pyramid since the march of Caesar's legions, was still a normally functioning, habitable ecosystem. But in the semiarid lands of the American Southwest—in the valley of Blue River or in Black Canyon for example— a cumulative process of wastage had set in. "This wastage in the biotic organism," Leopold noted, "is similar to disease in an animal, except that it does not culminate in absolute death. The organism recovers, but at a low level of complexity and human habitability." The combined evidence of history and ecology tended to support one general deduction: "the less violent the man made changes, the greater the probability of successful readjustment in the pyramid."[38]

Leopold had begun his study of deer at a time when he was intensely concerned about the dissolution of the southwestern landscape. The ability to maintain shootable populations of deer, it had seemed then, would be one small test of man's ability to live in harmony with his environment. It had never occurred to him that deer, like cattle, might actually contribute to dissolution. His criterion of proper stocking was normal stocking—the largest number of individuals a given range could carry, with stability of the population and of the habitat simply assumed. Although he had been fascinated by diversity, it had seemed to him that the very diversity of the country was somehow related to its instability; and even though he had argued

38. "A Biotic View of Land," *Journal of Forestry*, 37:9 (Sept. 1939), 729.

for preserving at least a sample of each species of wildlife native to the Southwest, his canon of wildlife diversity had not included mountain lions and wolves.

By the time he returned from the Sierra Madre of northern Chihuahua and began work at Huron Mountain, he understood that the absence of deer irruptions, such as had been responsible for the dissolution of Black Canyon and the Kaibab, was related in part to the presence of wolves and lions, which helped keep the deer herd in a productive, healthy equilibrium below the absolute carrying capacity of the habitat, that the centuries-long stability, or health, of the land organism was integrally related to its evolutionary diversity. The fundamental values, diversity and stability, remained, but the romance of diversity and the assumption of stability had been transmuted into criteria of environmental health.

5

Too Many Deer

The Public Problem

In his personal evolution toward an ecological attitude Aldo Leopold reached maturity during the 1930s. The public now had to be brought along the same route. At the start of his career, as we have seen, Leopold had taken his cause to the public and was notably successful in persuading them that game supplies could be restored by relatively simple measures, such as law enforcement, predator control, and refuges. As he became interested in more sophisticated environmental controls, he became acutely conscious of the need for professionalization of game management; hence he centered his activity during the 1930s on scientific research and the training of professionals. It was during these same years that Leopold began redefining objectives, symbolized in the profession by a shift from the economic term *game* to the more esthetic or ecological *wildlife*. Heartened by the receptivity to ecologically based values on the part of private landowners at Rockford and Huron Mountain, Leopold now challenged professional wildlife managers to concentrate their efforts on building broad public awareness through education and through ecological restoration on private lands. It would be long and slow, this process of changing ideas about what land is for, but it was the only way.

By about 1940, Leopold was probably prepared to devote the rest of his life to the slow, quiet process of building perception in human minds. To a large extent he did just that, as has been attested to by scores of his students, both liberal and professional, by farmers and other landowners with whom he worked, and by several generations of readers of *A Sand County Almanac*. But what he probably did not anticipate was the extent to which he would again become involved in the politics of conservation, this time in Wisconsin, much as he had been involved in New

Mexico during the game protection campaigns of 1915–
1921. He had entered the public arena in New Mexico in
response to a sense of crisis, when it seemed that game
animals, principally deer, were doomed to utter depletion
if action were not taken immediately. He got his message
across and led the public in action that appeared to allevi-
ate the crisis. Two decades later it was again a sense of
crisis that drew him into the public arena, only this time
the problem was that there were too many deer rather
than too few, a situation that would hardly have been
recognized as a crisis unless one had a rudimentary understanding of ecology. In a time of crisis, action may seem
more imperative than understanding, but in the long run
will not be tolerated if it goes contrary to public understanding. Leopold sensed this; but during the decade of
the 1940s he struggled simultaneously with the imperatives
of both action and understanding and learned thereby
what it meant to deal with an ecological issue in the public arena.

As the decade opened, he was involved in frequent correspondence with P. S. Lovejoy, former chief of the game
division of the Michigan Conservation Department, who
was in failing health but keen of mind and intent on
probing the history of ideas about land use. Leopold was
his sounding board. To Lovejoy, it seemed that it ought to
be possible to graph on a time scale the major nodes in the
sequence of human thought and action on conservation
problems, much as ecologists had learned to identify and
predict various stages in the succession of plant and animal communities for particular types of sites. Knowledge
of such a predictable sequence might make it possible, by
a process of 'ecological engineering', to shorten the interval between emergence of a problem, such as excess deer,
and public acceptance of action toward an effective solution. Ecological engineering to him meant recognizing
that man was still an animal and, as such, was not entirely
persuaded by reason or science. As Leopold paraphrased
Lovejoy's notion, "Reason, to the mass mind, is like oxy-
gen to the animal body: a little is essential, but too much is
toxic, and induces pain followed by defensive reactions."
Tolerance for reason could be increased by education, but

only by slow degrees. To bridge the gap between ecological research and public action therefore required various sorts of economic and social as well as scientific persuasions, or as Lovejoy expressed it, "placing another banana peel where it will do the most good."[1]

In a letter to Leopold in March 1941, Lovejoy worked out the sequence of ideas and action for the history of deer management in Michigan, where deer problems had developed and were recognized years ahead of Wisconsin. Seventeen years had already lapsed, he calculated, from first talk of "starvation . . . haul hay" to first sportsman-support in asking the legislature to "git modern in re deer." Within that interval, it had taken five years from talk of dead deer to the first postmortems; three years of digressions into live-trapping, mercy bullets, and hauled hay; another three years to begin browse surveys in deer yards, sex and age tallies (for the truth about "dry does"), and CCC census drives; two more years before the commission and department stated publicly that the only alternative to wholesale starvation was a higher legal kill, and another two years before they requested the legislature to modify the one-buck law; then two years after publication of an educational bulletin about whitetails, a deeryard movie, and public tours of deer yards until organized sportsmen began to support the request of the commission for authority to regulate deer by management units. Michigan had progressed to this point by 1941, at which time excess deer had not yet become a public issue in Wisconsin. Lovejoy then predicted that it would require another four to six years to get legislative authorization and five to ten years from first experimental efforts to partially successful "via-rifle regulation" over most of the state. Beyond that, he hoped that ten to twenty-five years might bring some measure of official recognition that

1. Leopold's letters from P. S. Lovejoy are in LP 1B2. From these he drew most of the material for his "Obituary: P. S. Lovejoy," *Journal of Wildlife Management,* 7:1 (Jan. 1943), 125–28, the source of the quotations on the following pages. Leopold did not necessarily subscribe in toto to Lovejoy's ideas, especially his notion of "ecological engineering" and his view of human nature, but as the primary recipient of Lovejoy's letters he felt an obligation after Lovejoy's death to give the ideas a hearing.

environmental management, including controlled burning and coordination with forestry, was even more essential than regulated hunting. He would not venture a guess as to how many years it would take after such recognition until policy had been "settled," workable techniques demonstrated, and the integration of forestry and deer management became a reality.

P. S. Lovejoy died within a year. In an obituary tribute to his originality as a "sire" of ideas about men and land, Leopold focused on his notion of ecological engineering and his theory of idea succession. Analyzing Lovejoy's essentially linear sequence of ideas on the Michigan deer problem, Leopold revealed his own profoundly ecological understanding of how ideas evolve. Lovejoy, he suggested, perhaps regarded the sequence of events before the population was reduced as comprising a single stage in the intellectual succession:

> Through it runs one idea: too many deer. It grew up under the dominance of a preceding idea (too few deer) just as the forb stage grows up under the dominance of weeds, or the grass stage grows up under the dominance of forbs. It will be succeeded by another idea: regulated deer, locally adjusted to the needs and tolerances of forestry and other land-uses. Too few deer, too many deer, and an adjusted deer herd are three overlapping stages in a succession.

This conception of the succession of configurations of ideas, rather than simply of discrete ideas, may shed light on the paradox that confronts us as we view Leopold's almost total preoccupation with the necessity of herd reduction in the 1940s, as against his broad-minded philosophical approach of the late 1930s. The five-step sequence of ideas leading from restriction of hunting to environmental controls, through which Leopold himself had passed by the time he defined it in *Game Management* in 1933, was actually a single stage in the intellectual succession. The dominant idea was too little game, or too few deer. Growing up under this was the idea of too many deer, which finally became dominant in his own thinking around 1936. But this idea was quickly succeeded by another idea that had also been germinating in his mind for some time—the idea of land health, which implied an adjusted deer herd. Having himself come to an ecological understanding of deer–wolf–forest interrelationships through a lifetime of thought and experience, Leopold during the decade of the 1940s would feel very keenly a responsibility to help bring the public through the same intellectual succession. The public was still thinking "too few deer"; they had been engineered for two decades or more to support the one-buck law and vigorous enforcement, wolf and coyote bounties, ample refuges, and even artificial feeding. To bring about a 180° change to the notion of "too many deer" would require more than a few strategically placed banana peels. And only after the public finally acquiesced in herd reduction could they be brought along a more difficult path to appreciate all the ecological implications of an adjusted herd.

Aldo Leopold was not by nature an engineer; he was much more comfortable in the role of educator or philosopher, and might well have preferred to look beyond herd reduction to broader considerations of land health. But in a crisis that was not yet recognized as a crisis by policymakers, let alone by field personnel or the media, who was going to assume leadership if Leopold did not? As P. S. Lovejoy put it, in his inimitable Lovejoyeese:

> Ain't it th helluva note that us Tops gotta come down offn high-cold peaks etc, for ta sweep out under beds & suchlike mean little chores, on acct of becuz them Social-sciencers aint even yet found out about Homo-sap-is-a-mammal & so framed etc?
>
> I dunno as I *want* to bother with being patient & sticking around the way Charlie Darwin done it. . . . waiting for the Widder of Windsor to git her arteries hardened up enough to suit Bishop Usher etc. How about you?[2]

Forebodings

The first systematic study of the deer situation in Wisconsin was finally initiated in the winter of 1940–1941,

2. P. S. Lovejoy to AL, May 1941, LP 1B2 (not quoted in the "Obituary").

under funding from the Pittman–Robertson Act for Federal Aid in Wildlife Restoration. The "P–R" program allocated money from an excise tax on sporting arms and ammunition to the states on a 75–25 per cent matching basis for habitat restoration and research. It had been enacted in 1937, and Leopold had hailed it at that time as the first federal recognition that stronger state game departments and state-based research programs were indispensable. This was especially true when it came to dealing with the problem of overstocked areas. "As long as the public retains its present almost incredible ignorance of range ecology," Leopold had written, "it is often immune to evidence, except such as is adduced by its own local representative."[3]

It was another problem to persuade the Wisconsin Conservation Commission to use Pittman-Robertson funds for deer research, however. Despite Leopold's urging, the commission dragged its feet for years, then decided its first priority was land acquisition for waterfowl in Horicon Marsh. Only in June 1940 when it became clear that the Horicon project could not meet federal standards by the end of the month, when the state's first installment of P–R funds (available since 1938) would have been lost, did the commission authorize a deer research project. There is a clear implication in the commission's biennial report for 1939–1940 that if options on the Horicon lands could have been secured in time, the entire P–R appropriation would have gone to that project, and projects to research deer and other game would not have been submitted for funding. Up to that time, most of the department's meager research efforts had been conducted at the Poynette game farm and had dealt almost entirely with wildlife diseases and maintenance of animals in captivity. Of twenty experimental projects listed in the biennial report for 1939–1940, the last three related to deer: analysis of deer antlers —to determine the chemical constituents of antlers so that the department would be "better able to answer the everyday questions of the layity" as to why certain rodents eat antlers in the wild; the analysis of deer milk—to determine

3. "Federal Aid for Wildlife," typescript, 10 Nov. 1937, 3 pp., LP 2B8.

how to fortify ordinary cow's milk for feeding to fawns at the state game farm; and deer castration—to determine whether castration would give bucks a gentler disposition and inhibit antler development, which "should be of benefit to the deer farmer as it eliminates the hazard of holding vicious bucks." Small wonder that Leopold tended to be critical of the department's sense of priorities or that his approach to game management should have seemed such a radical departure.[4]

The new project on deer was headed by William Feeney, a biologist from the conservation department who had worked on the University of Wisconsin Arboretum in the 1930s while Leopold was director of animal research there. Leopold was apparently asked to sit in as a continuing advisor. First priority for Feeney and his crew of about a half-dozen men was to undertake a survey of browse conditions in the winter yarding areas in the northern part of the state. Indeed, during most of the 1940s the work of the Feeney crew could be described more properly as survey than as research.

The workers on the deer project, after their first winter on the job (1940–1941), reported that in 91 per cent of the eighty-one deer yards examined cedar was heavily browsed or completely browsed out, and balsam, at that time regarded as a starvation food, was browsed in 75 per cent of the yards. Starved deer were found in over a third of the yards, and the crew estimated that more than half the fawn population died of starvation in some of the more heavily browsed yards that winter.[5]

4. Wisconsin Conservation Commission, "Seventeenth Biennial Report," 1939–1940, manuscript (Madison, 1941), pp. 83–84.

5. The deer project was designated Pittman–Robertson Project W–4–R. Mimeographed quarterly progress reports, ranging in length from six to about thirty pages, were regularly prepared by the project leader and issued by the conservation department, as required by federal law. A complete set may be found at the research library, Nevin State Fish Hatchery, Madison. The first report is dated April 1, 1941. Ernest Swift summarized the 1941–1944 reports in his *History of Wisconsin Deer*, Wisconsin Conservation Department Pub. 323, March 1946. Burton L. Dahlberg, who took over as project leader in 1948, and Ralph C. Guettinger, who became leader in 1950,

Too Many Deer

With this type of data, one could begin to demonstrate the reality and the magnitude of a deer problem in Wisconsin. No longer could overbrowsing and starvation be viewed as a localized condition; it was generalized over at least the northern third of the state. Ahead was the task of persuading public officials and citizens that more than winter feeding was required—in short, that there were too many deer.

The Challenge of the Kaibab

The magnitude of the problem and the imperative need for courageous leadership in dealing with the long-term consequences of too many deer was driven home to Leopold the following summer when, for what was apparently the first and only time in his life, he visited the fabled deer ranges of the Kaibab National Forest. This trip, a tour of cooperative wildlife research units that he undertook as chairman of the American Wildlife Institute's technical committee, may have been his first direct field examination of severely overbrowsed deer ranges anywhere in the West. More than fifteen years had elapsed since the starvation winters of the mid-1920s, and one might have expected most signs of the earlier debacle to be obliterated. Indeed, administrators on the Kaibab believed that the range was well-enough recovered to start rebuilding the herd, and they had already reinstituted official predator control for lions and coyotes. Leopold disagreed. The objective, he asserted, should be to rebuild the original richness of the range rather than merely retain a tolerable condition: "Hence, the true yardstick of progress should be the establishment of *new reproduction of the palatable browse species* to replace the dead stems lost during the irruption." Mere recovery of stems that had managed to survive the irruption was not satisfactory.[6]

Although he found many national forests, parks, and

___
wrote the official report of the project: *The White-Tailed Deer in Wisconsin*, WCD Technical Wildlife Bull. No. 14, 1956.

6. "Report to American Wildlife Institute on the Utah and Oregon Wildlife Units," typescript, 10 Aug. 1941, 14 pp., LP 2B1.

refuges in even worse condition than the Kaibab, Leopold was amazed to find that some ranges still in fair condition were capable of carrying enormous deer populations, far above any the best eastern deer yards could boast. An enclosure in the Fish Lake National Forest in Utah, for example, was carrying 300 deer on 700 acres yearlong. "This carrying capacity of a deer per two acres," he wrote, "must modify all our thinking on the deer problem." The discovery that the browse ranges of the Great Basin were not impoverished semidesert scrublands, but rather "an extremely rich national resource," gave special urgency, in his thinking, to the predator question. The common characteristic of all the overbrowsed deer ranges he knew about was that predators had been eradicated, and yet "in the face of these almost obvious lessons" operations to control predators were being continued without protest. The common assumption was that rifles would do the necessary trimming, but that was naive, Leopold charged, for rifles had never yet trimmed enough deer in time to save a threatened range: "Hunting is a crude, slow, and inaccurate tool, which needs to be supplemented by a precision instrument. The natural aggregation of lions and other predators on an overstocked range, and their natural dispersion from an understocked one, is the only precision instrument known to deer-management."[7]

Leopold challenged local researchers and administrators to take a stand on issues of range quality and natural predation, but he was dismayed to find that most of them were satisfied with token recovery of browse and willing to acquiesce in predator eradication. In an article on a related issue, the large-scale invasion of overgrazed western ranges by an inferior annual weed known as *cheat*, he commented on the tendency he had observed among local citizens, including technically trained range and wildlife managers, to regard such problems as necessary evils "to be lived with until kingdom come" rather than as chal-

7. That Leopold and others may have overstated the case for predators on the Kaibab is suggested by Graeme Caughley, "Eruption of Ungulate Populations, With Emphasis on Himalayan Thar in New Zealand," *Ecology*, 51:1 (Winter 1970), 53–72. For further discussion of the issue of predation see pp. 209–17.

lenges to rectify past errors in land use. He was reminded once again how difficult a task confronted the coming generation of technical men:

> How to join the life of a local community without "going native" intellectually; how to muster courage to unravel land-use problems which are, at best, only partly soluble; how to become expert in one small technical field without losing the common touch with land as a whole; how to translate technical knowledge into land-use practice without loading the whole job on the government; these are tasks indeed![8]

After his trip to the West, Leopold set about trying to organize and find funds for a cooperative study of what he was now calling deer *irruptions*. "I can't see why it is any more of an 'Act of God' for the National Forests to be chewed up than to be burned up. I suspect you can't see why either," Leopold wrote in his plea for funding to a long-time friend Earle Clapp, who was Acting Chief of the Forest Service. "Perhaps this project offers a way to shake the Western technicians out of their mental grooves."[9]

The plan was for Richard Costley, a young range manager with the Utah district of the Forest Service, to try to deduce the nature of the irruption mechanism by comparing unhealthy with healthy herds. By health Leopold meant "that capacity for self-adjustment in a population which insures against both over-population and under-population." If Costley were going to study a healthy herd, he would have to go south of the border. "My trip through the West this summer further convinced me of a fact which I have long suspected," Leopold wrote Costley:

> There are no really healthy deer herds left in the United States. You can find on this side of the border examples of

8. "Cheat Takes Over," *The Land*, 1:4 (Autumn 1941), 310–13. Mature cheat grass with its prickly awns was not only inedible to livestock but also highly inflammable. Cheat fires in the foothills burned out the remnants of good browse such as bitterbrush, sage, and oak in the narrow wintering belt of the deer.

9. AL to Clapp, 18 Dec. 1941, LP 2B2.

almost any kind of ecological distortion, but I cannot think of any herd which I could conscientiously call normal.

Leopold had never lost his desire to study a normal herd, a desire dating back at least to the early 1920s and the Gila. But by normal he now meant "healthy" rather than "best-stocked," and the only healthy herd he knew was along the Río Gavilán in the Sierra Madre of Chihuahua.[10]

As for the comparison with abnormal herds, it was Leopold's thought that Costley might analyze data already on hand for various irruptive herds in the United States to look for some common denominator that "might give us presumptive evidence of causation." Such a common denominator, as he suggested repeatedly in correspondence and in print, might be the eradication of predators. Thus on the one hand he seemed to be looking for a single simple "cause" of irruptions, while on the other hand he sought a deeper understanding of complex population mechanisms. It was the latter, more ecologically sound approach that won out when he actually formulated a preliminary hypothesis for a deer irruption study.

Having been much impressed with the high percentage of palatable browse species available to deer in the Great Basin, he grounded his hypothesis in the accumulation of huge browse reserves on ranges originally kept "nearly deerless" by predators and, when the area was first settled, by heavy, unrestricted hunting. He compared these browse "impoundments" to alluviating watersheds:

> Such impoundments approach unstable equilibrium in respect of deer, just as alluviating watersheds approach unstable equilibrium in respect of erosion. The first accidental relaxation of predator-pressure brings on the irruption: exhaustion : die-off sequence recently illustrated on the Kaibab. Just so combinations of drouth and rainfall precipitated the erosion cycle in vulnerable impoundments of soil.

10. AL to Costley, 13 Sept. 1941, LP 2B2; see also, "Wilderness as a Land Laboratory," *The Living Wilderness*, no. 6 (July 1941), 3.

Deer irruptions, like stream erosion, occurred throughout presettlement times but "active periods were self-terminating," Leopold suggested, presumably because excess deer induced increased predator activity. Overgrazing, lumbering, and fire control after white settlement contributed still more to the browse impoundments. Then came hunting laws and strict enforcement, eradication of predators, and creation of refuges, which "acted in various combinations as the 'trigger-pull' to activate all vulnerable deer ranges almost simultaneously." These man-induced irruptions were not self-terminating so long as predator control, one-buck laws, and overgrazing continued.[11]

There is at least the implication in Leopold's hypothesis that deer irruptions had their own internal dynamic or, as he put it, that they were "self-aggravating." He believed there were indications that heavy browsing by irruptive herds might initially stimulate greater quantity and better nutritive quality of browse, which in turn might increase reproductive rates and perhaps even distort the sex ratio toward females, further increasing the size of the herd. There is also the suggestion that ranges such as those in Utah and the Kaibab with a high percentage of palatables in the woody vegetation might be particularly vulnerable to violent irruptions:

> Palatable species comprise the bulk of the winter range, whereas in the Lake States, the Arizona brushfields, California, Oregon, and Pennsylvania they comprise but a small fraction. Is this inherent or induced? Is a "pure" range, once upset, more liable to violent ups and downs than a "dilute" range offering inferior "buffer" foods? Presumably yes, especially if overbrowsing enhances quality.

Leopold obviously discovered some promising leads for fundamental research. Most important was his emphasis on a possible relationship between reproductive rates and the nutritional quality of browse, which has been well established by subsequent research.[12] But Leopold was not

11. "Deer Irruption Study," typescript, 18 Dec. 1941, 6 pp., LP 2B2.

12. See for example A. Starker Leopold, et al., *The Jaw-*

to have the benefit of such research findings as he struggled with the deer problem in the remaining years of his life. Pearl Harbor intervened, funding became unavailable, and Costley went off to war.

Selling a New Idea

Commenting on the possible relevance for the Wisconsin deer project of Leopold's irruption hypothesis, especially with respect to the seemingly less violent oscillations on the more "dilute" ranges of the Lake States and the vagaries of public opinion, William Feeney had a startling thought. "Could it be that if we didn't try to check the compounding upward irruption of Wisconsin deer, both the deer and the forest would be better off?" he queried.

If the herd were more suddenly reduced to near zero by an extremely hard winter when the population was at a peak it might be better than a prolonged struggle of mediocracy which could result from far lagging artificial controls pre-climaxly applied. In other words, is there a possibility of a deer herd remaining just large enough to eat all of the

---

bone Deer Herd, California Division of Fish and Game, Game Bull. No. 4 (1951), 110–12; Richard D. Taber, "Deer Nutrition and Population Dynamics," *Transactions*, Twenty-first North American Wildlife Conference (March 1956), 159–72; and summaries of other research in W. P. Taylor, ed., *The Deer of North America* (Harrisburg, Pa., 1956), William Dasmann, *If Deer Are to Survive* (Harrisburg, Pa., 1971), and Aaron N. Moen, *Wildlife Ecology* (San Francisco, 1973). In general the research indicates that nutritive quality of browse (as signified for example by protein levels) varies significantly not only from species to species but from area to area (better soils and recent burns producing better browse) and from month to month (moisture, sunlight, temperature, etc.); and perhaps even more significant in terms of Leopold's hypothesis, that heavy browsing of certain shrub species promotes greater quantity and also higher moisture and protein content of regrowth. Leopold, however, apparently thought more in terms of vitamins than of protein. Ovulation rate, fawn survival, and general weight and vigor of deer have been correlated directly with variations in nutritive quality of browse. Physiological studies of deer also reveal variations in nutritive requirements depending on sex, age, weight, season of year, and other factors.

conifer reproduction until the seed trees have fallen? If so, we could lose both the vulnerable forest and the deer in the northern belt—imaginative to be sure but chances are it is a tendency worth some speculation.

Leopold's penciled notation in the margin: "Good." "Of course we would like to try management," Feeney added as if realizing suddenly how radical his suggestion might sound, "especially unhampered and well thought out management—and there is still plenty of thinking to do."[13]

Feeney no less than Leopold was beginning to realize what obstacles they were going to face in Wisconsin, especially since there was as yet no reason to hope that management would be unhampered even if it were well thought out. Right from the start, Feeney recommended substantial reduction of the herd and regulation of logging in the wintering yards. And from the beginning the reports and recommendations of the Pittman–Robertson project met with distrust and deep resentment, not only by the public but also by local wardens, rangers, and other members of the department's vast field force.

Although some of the earliest reports of deer damage and starvation had come from wardens and rangers and several of these individuals cooperated fully with the P-R crew, the enforcement men as a group were still predisposed to think in terms of protecting deer from illegal kill in order to build up the herd, and they were proud of what they had accomplished. True, deer died in certain isolated areas during severe winters, but a certain amount of starvation was to be expected and could be allayed, many felt, by feeding in the worst problem areas, as had been the practice since 1935. Men who have devoted their energies to a cause, whether it be hauling hay to starving deer in subzero cold or apprehending poachers, or something more, and especially if it be voluntary service, want to believe that what they did was worthwhile. To suggest, as both Leopold and Feeney were wont to do, that feeding deer did more harm than good and that the illegal kill, estimated as high as 40–50 per cent of the legal kill, was

13. W.S.F. [William S. Feeney], "Comments on 'Deer Irruption Study,'" typescript, 3 Dec. 1941, 2 pp., LP 2B2.

actually a blessing in disguise, was not calculated to win friends among the field force or other concerned citizens of the north country.

Wardens and rangers in those days were not, for the most part, technically trained men. Then, too, overbrowsing was an insidious thing. If a highly palatable species were being replaced by one less palatable, how was one to know that it was happening, or what was causing it, or even what this thing called *palatability* was? "Now come these young theorists from Madison who know nothing about the north country, and by their own admission cruised much less than 1 per cent of the area (1941), and *they* try to tell *us* that we have too many deer," was the attitude of many wardens and rangers. Except for a few parasitologists, pathologists, and veterinarians cloistered at the Poynette game farm, the conservation department had had no active wildlife research programs. Hence many wardens had never encountered a wildlife biologist in the field and had grown accustomed to thinking of themselves as the ranking experts.

Feeney, in turn, did not have much use for the field force. At a meeting of division chiefs and field supervisors in 1942, the chief conservation warden asked whether Feeney's data were based on all available information. Yes, said Feeney, on the information provided by his own nine honest men. The warden wanted to know whether Feeney had contacted the field force, and if not, why not. Feeney's reply: "Sometimes it helps to contact them and sometimes it is a headache, to be honest with you." He admitted this reply showed prejudice, but added that the information given by too many of them was not worth anything, so contacting the field force systematically was a waste of time. It is easy to understand the resentment of many in the field force that department officials in Madison should take the word of Feeney's research crew over that of long-time employees who were, in their own estimation, intimately acquainted with and responsible for their own areas. It is also easy to see how public distrust of the research findings and antagonism toward the department in Madison could feed on this resentment felt by wardens and rangers who were, after all, members of their local communities.[14]

In time the conservation commission would realize the wisdom of involving wardens and rangers in the deer yard surveys, but their first response to widespread distrust of the research findings and charges of secretiveness regarding the Feeney survey was to appoint a Citizens' Deer Committee, in September 1942. Members of the committee were asked to make an independent investigation of the situation, both as individuals and as a group, and to involve as many other interested citizens as possible. Aldo Leopold was selected as chairman of the nine-member group—whether by the members themselves or by the chairman of the commission is uncertain. There is no evidence that the committee was loaded in any way, except with a cross section of interested citizens mostly from the north, but someone would have to have known that Leopold, who was especially effective in a committee situation, would take the ball and run with it, and in Feeney's direction.

To provide a broader context and scientific support for the work of the citizens' deer committee, Leopold put on his hat as chairman of the natural resources committee of the Wisconsin Academy of Sciences, Arts, and Letters to prepare a historical summary of the deer problem nationwide. In it he traced the irruption history of four herds whose fortunes he had been following for over a decade— the Kaibab (Figure 3), the George Reserve in Michigan, the lower peninsula of Michigan, and Pennsylvania. It

14. See verbatim transcript of deer discussion, Tomahawk, 30 June–1 July 1942, in WCD files, SHSW Archives. Feeney's personality may have been a factor in the deer problem. Although professionally capable and sincere, he apparently had an abrasive effect on his staff and others both in and out of the department and was ultimately forced to resign. A more subdued personality conceivably could have resulted in a quite different reception for the early deer research project's findings. The commission did not help matters either. Minutes of 11 Aug. 1942 indicate that the commission decided in executive session that no publicity be given to Feeney's report on the present status of Wisconsin deer, at least for the time being.

is worth noting that he made no mention in this report, nor in any other public statement during the 1940s, of his personal involvement with game protective activities and deer research in the Southwest or of the irruption in Black Canyon. Comparing the history of the various herds, Leopold described what he termed an *irruption sequence*. In the first stage, a buck law and enforcement, predator control, and refuges, often in conjunction with wide-spread logging and some fire, allow the herd to multiply. A deer-line appears on palatable browse, but the herd continues to increase. Then a deer-line appears on unpalatable browse, and fawns die of starvation in hard winters. The herd is at its peak. Unless the herd is sufficiently reduced at this point, adults begin to die, the herd is down-graded by disease and shows light weight and small antlers, and palatable browse begins to die off, only to be replaced by unpalatable species. Eventually a new equilibrium is reached, but at a level well below the preirruption carrying capacity of the range. The Kaibab herd had declined to only 10 per cent of its peak level, Leopold pointed out, and in Pennsylvania, where doe seasons were instituted but not readily accepted by many people, the herd was reduced by more than 50 per cent and still exceeded the carrying capacity.[15]

15. Leopold's file on the Wisconsin Academy committee is in LP 2B10 and lecture notes for preliminary presentations of the irruption histories are in LP 6B14. The full report was compiled by Leopold for the natural resources committee, which also included Ernest F. Bean and Norman C. Fassett; it was published as "Deer Irruptions," *WCB*, 8:8 (Aug. 1943), 1–11, and reprinted in Wisconsin Academy of Sciences, Arts, and Letters, *Transactions*, 35 (1943), 351–66. For further analysis of the irruption hypothesis, see pp. 241–47.

Leopold also published a brief article, "The Excess Deer Problem," *Audubon Magazine*, 45:3 (May–June 1943), 156–57. He had been asked by the executive director of the National Audubon Society to write an article criticizing people who were advocating increased hunting of wild game in order to provide additional meat during World War II and to control surplus animal populations. Leopold took issue with Audubon instead, and wrote an article about deer irruptions in which he argued for prompt reduction by means of both hunting and predators. See John H. Baker to AL, 17 Dec. 1942, and AL to Baker, 23 Dec. 1942, LP 2B5.

Too Many Deer

Leopold presented a preliminary version of the histories of the deer irruptions as a slide lecture at the first meeting of the citizens' deer committee in January 1943. But he was upstaged by another member of the committee intent on presenting a different view. Joyce Larkin, editor of the *Vilas County News-Review*, who had been without a doubt the sharpest thorn in the side of Bill Feeney and his deer research crew, appeared at the first meeting armed with a printed "book" of history and opinions on the deer situation. In it she told how Vilas County, which had been subjected to the heaviest buck kill of any county in the state back in 1936, had been violently opposed to an open season in 1937 (the first open deer season in an odd-numbered year since 1923) and how the county board, men's clubs, and sportsmen had cooperated to build "a sort of psychological fence" around the county in order to discourage outsiders from hunting there. For the next five years the delegates from Vilas County at the annual game hearings voted consistently against open deer seasons, and local public sentiment mounted against the conservation department, which just as consistently turned a deaf ear. When Joyce Larkin invited all her Vilas County brethren to write or tell her their views on the deer question in the fall of 1942, she was inundated with responses.

Of all the people who came to her office, most of them woodsmen, resort owners, and guides, "not one of them," she emphasized, "has ever seen a starved deer. They have seen dead deer." These men thought deer were dying from old age, pneumonia, disease, and especially from crippling during hunting season. They regarded the herd as greatly depleted and unbalanced, with small bucks and many too many "dry" does, which were eating browse and not producing offspring—shades of Aldo Leopold's analysis of the Gila herd in 1929. Because Vilas County had been a cutover area, it could support a larger herd both summer and winter even without cedar browse, they thought, especially if logging operations were properly spaced and timed. But in winters of deep snow, like the current winter (1942–1943), deer had trouble moving about to reach food, and it might be necessary to make snowshoe or tractor trails for them and knock off some of the higher cedar branches to make browse accessible. All but two of the men she had interviewed favored a closed season on deer for from two to five years.[16]

It should be recognized that Vilas County was a resort area in the heart of northern Wisconsin's lake country. In the 1940s the tourist business was strictly a summer affair, and the chance to see deer, not to shoot them, was regarded as a major tourist attraction. County residents would "take" their deer anyway, if they were so inclined, but it hurt them sorely to see their bucks leaving the county on the fenders of automobiles from downstate. Joyce Larkin herself had helped create such attitudes. She had arrived in the area in the early 1930s, a time of flagrant poaching and indulgent courts, with the intention of establishing a conservation newspaper; and she got by, she told Leopold some years after their first meeting, only because she was a woman and people figured she just didn't know any better. As she related her exploits in changing public attitudes:

> My first step in selling conservation was to hammer home the value of the tourist industry, how much in dollars and cents it brought to this region, how it supported an entire population, and then I pounded away at why the tourists came into the region, because of the woods, waters and wildlife. When I was sure that the relationship between the tourist industry and the wild life was established so that even the worst tobacco-spitting trapper who hadn't had a new idea in the last 20 years was convinced, I started on the enforcement angle.[17]

Thus, by the time she began serving on the citizens' deer committee she had public opinion solidly behind her on the need to protect the deer. It is probably safe to say that

16. Joyce M. Larkin, "A Report on the Deer Situation in Vilas County," 23 Jan. 1943, WCD files, SHSW Archives. The argument that deer did not die of starvation was not without foundation, since the direct cause of death was usually pneumonia or other pathological problems induced by malnutrition. Insufficient or inadequate food was indirectly the cause.

17. Larkin to AL, 31 Dec. 1944, files of Walter E. Scott.

most other members of the committee, as of the first meeting, shared her views.

It was an ironic bit of fortune that the winter of 1942–1943 was a particularly severe one in Wisconsin, and deer starved by the thousands. Having a monumental selling job to do, Leopold arranged a three-day tour of eight northern yards for his committee. With an itinerary prepared in advance, invitations to members of the department field force and local conservation groups, and notice to newspapers throughout the state, the tours attracted about sixty people a day, most of whom went out into the yards on snowshoes. They saw about a hundred dead fawns and watched as several others were caught by hand and died. That there was extensive browse damage to the best food species and that deer were eating inferior foods was readily apparent to all when pointed out by the experts; that dead deer had starved was clear from autopsies performed on the spot, which revealed paunches full of inferior foods like balsam and bone marrow that was reddish, translucent, Jell-o–like—an indication that the last reserve of fat had been used up.

One of the reporters present on the tour was Gordon MacQuarrie, long-time outdoor writer for the *Milwaukee Journal*, the most influential newspaper in the state. MacQuarrie regarded Leopold as "the greatest news tipster" of his experience. "He was alive with ideas," said MacQuarrie, "and it should be added that he knew news when he saw it, and knew what to do with it." The deer-yard tours were news. The following is from MacQuarrie's description of the tour in the Flagg Yard in Bayfield County, one of the biggest and worst-hit yards of them all, in which overbrowsing had been reported since the mid-1930s. Leopold's group has just come upon two more deer:

> One falls, unable to move. The other runs a few yards, is easily captured. It is so thin that Buss says "you could shave your face with his backbone." He's a little buck, nobs barely protruding on his head.
>
> Nasty story? Ask Mrs. Harry Thomas of Sheboygan, one of the committee, who went out with the rest through the

snow. Ask her if she was able to take it when husky young Mr. Buss picked up that dying baby deer and, already in a coma of death, it tried to raise its head and look about. And if you ask her, tell her there were those among the throng who saw that she couldn't take it, because deer are not supposed to die the hard way.

Nasty story? Every whipstitch of that trip through the Flagg yards was that and worse—

Dr. Chaddock, the pathologist, pointing out with professional skill the secondary invaders that moved into the carcasses when the food ran out. Dr. Aldo Leopold, University of Wisconsin wild life research director, muttering, "Same thing in Utah, in the Kaibab forest; too many deer for the available feed." Virgil Dickinsen, conservation commissioner, remarking, "Whatever we do about it they'll crucify us."[18]

As a result of his experience with the tours, Leopold became convinced even more firmly that the only way to get people to understand the problem was to take them into the yards as he had done and show them. But there was a limit to how many people he could reach that way. MacQuarrie commented on the situation:

Up in that Flagg River deer yard two weeks ago where a mere handful of citizens were shown wholesale death by starvation among the deer, Aldo Leopold, trudging along on his snowshoes, spoke the old, old truth about conservation management. Resting a minute while a post-mortem was performed on a dead whitetail he said:

"The real problem is not how we shall handle the deer in this emergency. The real problem is one of human management. Wild life management is comparatively easy; human management difficult."[19]

That visiting the right yards with the right people was all-important in the deer business is indicated by a letter to Leopold from Dr. E. G. Ovitz, a committee member

18. "Here Come the Biologists," *Wisconsin Academy Review*, 6:4 (Fall, 1959), 159; "Death Stalks Deer Country," *Milwaukee Journal*, 23 March 1943.

19. "Prejudice or Science: That is the Issue," *Milwaukee Journal*, 4 April 1943.

who had missed the trip. Dr. Ovitz noted that in Forest and Oconto counties, which had not been visited at all by Leopold's tour, he himself had visited ten yards and had found none of them overbrowsed, except for two in the Argonne Refuge, a perennial trouble area. Local people were upset that Leopold's committee had not seen their area because they felt their herd was under control. "If the season is opened to kill antlerless deer next fall in Forest and Oconto Counties, without an investigation," Ovitz warned, "merry hell will be to pay."[20]

At a meeting of the citizens' deer committee some time after the tour of the deer yards, Joyce Larkin showed up again with a report on the Vilas situation. She wrote, "When I arrived at the first meeting of this group, I brought with me a booklet which contained what I now recognize as opinions. This time I have arrived with facts." Aldo Leopold had challenged local interests to assume more responsibility, and Vilas County, she was proud to report, had accepted the challenge:

> That the present deer problem consists of scores of local problems, all more or less different, is the opinion of Prof. Leopold. He feels local conservation officers and local residents should express local needs and assume partial responsibility for local policy. He thinks the commission might well encourage experimental local trials of various plans for bringing clashing interests together locally, rather than pressure groups at the capital.
>
> Vilas County has done just that. It has made a local survey and has united its people through the county board on a policy for which it will assume responsibility. Its county board has made a careful study of the local situation. It has facts to present, not theories, panaceas or sentimental and poetic outbursts.

After Leopold led his tour the county board had appointed five men, who surveyed a total of thirty-one more yards in the county. The men found thirty-two dead deer, a number of them starved. They found browse conditions spotty, in general better in the east and south and poorer

20. Dr. E. G. Ovitz to AL, 30 March 1943, WCD files, SHSW Archives.

in the west, but they thought that starvation and semi-starvation areas could in time develop in various regions if the herd continued to increase. "In Vilas county," wrote Larkin, "we are reasonable. It is true that the biggest outcry over a proposed antlerless deer season was expected to come from Vilas county, which has been accused unjustly of being out after the non-stop whining championship over dead deer." But Vilas County had made its own survey and, although county board members were glad that conditions were not as severe as in some other counties and in the Flagg yard, still they did not want to see things slide, and they had therefore agreed to go along with the state on an antlerless season. The board urgently requested, however, that the hunt be controlled, at least to the extent of limiting licenses to those who had taken out licenses in previous years. If such limitation were not enforced, especially in view of wartime meat rationing, Larkin warned, "everyone in a family, Pa, Ma, Aunt Ida, Grandma, Junior and Uncle Bill, will take out licenses and these licenses will be filled."[21]

Fascinated as she was by the process of creating public opinion, Larkin did not waste the opportunity provided by her report to teach the conservation department a thing or two about strategy. She was particularly dismayed with all the emphasis on starvation. The Feeney crew and a few converted local sportsmen had been going out into the yards after severe winter storms and piling up the carcasses of starved deer for photographing. The department had just produced a stark propaganda film titled *Starvation Stalks the Deer,* which its men were showing nightly in meeting halls around the state. But the public, Larkin pointed out, "doesn't understand starvation, and won't. It will either hysterically demand feeding programs that can't be carried out successfully or it will avert its eyes and insist deer do not starve." The department's approach was thus fruitless.

"To gain public understanding," she volunteered, "it is necessary to start with a fact that the public knows.

21. Larkin, "Comments on the Deer Situation in General and in Vilas County in Particular," typescript, n.d., 15 pp., WCD files, SHSW Archives.

That fact is present. It is that the herd is unbalanced. There are too many does and too few bucks. Every hunter will agree that this is true, because he has actually seen this condition in the woods and has annually complained about it." From the condition of unbalance stemmed most of the afflictions of the herd—immature bucks producing weak offspring unable to withstand the rigors of winter and excess dry does, unbred because of buck shortage. The obvious remedy was a season on does to redress the balance and allow the bucks to mature. Wasn't an antlerless season precisely what the department was after anyway, when they argued starvation? She was saying that in order to create public opinion it was necessary first to find some point of agreement and then build one's case from there, one idea at a time, much in the same manner as she had promoted conservation and law enforcement in Vilas County.

Aldo Leopold shared Joyce Larkin's dismay over the inadequacy of the department's public relations efforts and especially over the film *Starvation Stalks the Deer*, which he thought was misleading in its emotional play on starvation and in its lack of attention to the larger questions of range quality.[22] The barren doe–buck shortage argument advanced by Larkin as a substitute for the starvation issue, however, was representative of an earlier stage in the sequence of ideas about game management, as Leopold undoubtedly realized, having been over that route himself. He was beginning to appreciate that underdeveloped bucks and low fawn counts could be attributed to poor range quality as well as to hunter-induced buck shortage. Granted there was a superabundance of does, but the plain fact of the matter was there were also too many deer. To Leopold, the excess-doe argument was now different from what it had been around 1930. A doe season was required not because does were barren and unproductive, but precisely because they were productive. Since deer were polygamous, only by killing does could the future herd be reduced.

22. See for example AL to V. L. Dickinsen, 16 Oct. 1946, concerning the film, and other correspondence 1943–1946 in Leopold's conservation commission file, LP 2B10.

For whatever personal rationales, Aldo Leopold, Joyce Larkin, and the other members of the citizens' deer committee, most of whom had originally opposed the shooting of does, voted with only one dissent to recommend to the conservation commission an antlerless season for 1943. The committee report, drafted by Leopold, expressly recommended elimination of the usual buck season for three reasons: to redress the sex ratio; to demonstrate that herd reduction, which could be accomplished only by shooting does, was essential; and for considerations of safety (the idea being to force the hunter to look for horns or no horns before he shot). In addition to an antlerless season, the report called for more responsibility at the local level for identifying and meeting problem situations, for more objective measurements of deer damage, a system of fenced enclosures for comparison of unbrowsed with browsed vegetation, rescinding of oversized or overbrowsed refuges and other closed areas prior to the 1943 deer season, and an educational program to "teach citizens how animals and plants live together in a competitive-cooperative system." The committee also recommended that in the wilder counties "a low population of timber wolves be deliberately maintained as insurance against undue congestion or excessive numbers of deer."[23]

One member of the committee, Judge Asa K. Owen, published a minority report in the *Milwaukee Journal,* in which he argued against an antlerless season and in favor of winter feeding. In his estimation, the scientists had not demonstrated that there was an undue proportion of barren does, nor that there was any shortage of food outside of the yards. The very fact that deer concentrated in the yards during severe weather, he asserted, made an emergency feeding program feasible. "In other words," he wrote, "until there is a lot more known, the planned

23. "Majority Report of the Citizens' Deer Committee to Wisconsin Conservation Commission," *WCB*, 8:8 (Aug. 1943), 19–22; reprinted in "Wisconsin's Deer Problem," WCD Pub. 321, pp. 20–23. The report was strongly worded, with a good many loaded words like *wiped out, always, nullified, not one, denuded,* but it contained a tabulation of committee members' votes on particular questions at issue, which indicated that they subscribed almost entirely to Leopold's analysis.

reduction amounts to an experiment, and why experiment? A lot of good men and women have worked to build up a conservation sentiment in this state. The people generally, who are now interested, are not going to like this sudden reversal of policy, whatever the excuse for it." Owen was challenging the experts, and urging the people to think for themselves. His kind of argument was incredibly powerful, as Leopold would learn if he did not already know.[24]

Leopold reported orally to the conservation commission at its May meeting, although the written report of the citizens' deer committee was dated June 9, 1943. Early in June, Leopold and another member of the citizens' deer committee, John Moreland, an insurance agent from Hayward, were appointed by Acting Governor Walter Goodland to six-year terms on the conservation commission; their appointments were confirmed by the state senate on June 17, effective July 27.

Commissioner Leopold and the "Crime of '43"

Aldo Leopold's involvement with the state conservation commission and department in 1943 was a natural thing, but a long time in coming. After his initial involvement in establishing the commission in 1927, his relations with the agency progressively deteriorated, despite his best efforts. Then in 1939 his stalwart from the conservation battles of the 1920s, Attorney W. J. P. Aberg of Madison, was appointed to the commission, and Ernest Swift, one of the first to appreciate the gravity of Wisconsin's deer problem, was elevated to deputy director of the department. Thereafter, Leopold's relations with the state began improving.

His appointment to the commission was part of an effort by a new governor to escape from a political imbroglio that had engulfed the commission, the legislature, and several state agencies during the previous administration, and had led to the summary firing of H. W. MacKenzie (director of the conservation department), badly tarnished the department's image, and contributed as well to the

24. Asa K. Owen, *Milwaukee Journal*, 30 May 1943.

defeat of the incumbent Republican governor, Julius P. Heil, at the hands of a rival faction in his own party led by Leopold's close personal friend, Madison industrialist Tom Coleman, who became state Republican chairman.[25] Although Leopold was himself about as apolitical as a man in public life could be, his friendship with Coleman was almost surely a factor in his appointment to and acceptance of a seat on the commission. A number of Leopold's close friends urged him not to step into the political fray, arguing that a man of his temperament could be equally effective working from his academic chair—and at much less personal sacrifice. But he apparently felt that "a man ought" to accept such responsibilities in the public arena once in his life, and that his time was then.[26] His

25. The investigations were interpreted in some quarters as an attempt by Republicans and some Democrats under Heil to clean house after four years of "socialistic" administration under Philip La Follette, a Progressive. Wisconsin, traditionally a Republican state, had a three-party system during the 1930s and early 1940s, a situation that on the one hand gave conservation interests a certain amount of political leverage in elections and, on the other hand, injected more politics into the commission and department than they were designed to accommodate. Heil was defeated by the Progressive candidate, Orland S. Loomis, who died prior to his inauguration and was succeeded by Lieutenant Governor Goodland, a Republican.

From then until Gaylord Nelson's Democratic victory in 1958, Republicans would maintain control of the executive office, and the traditional north–south split would once again be the dominant factor in Wisconsin's conservation politics, especially when deer were at issue. Republicans were especially strong in the North, which had a disproportionately large representation in the legislature, the last major reapportionment having been in the 1890s at the height of the logging boom. This situation allowed little leverage within the party for ecologically minded individuals like Leopold in the southern part of the state.

H. W. MacKenzie was replaced as director of the conservation department by E. J. Vanderwall, who had been markedly successful in organizing the department's forest fire protection system but as director was almost a nonentity. Leadership on the deer question gravitated to the assistant director, Ernest Swift, who in 1947 was elevated to the directorship.

26. W. Noble Clark, emeritus director of the U.W. Agricultural Experiment Station, recalled how he and others had

appointment was hailed as entirely meritorious and non-political, a boon to conservation in Wisconsin.

After his appointment to the commission but before he actually took office, Leopold transmitted the report of his citizens' deer committee to the Wisconsin Conservation Congress, a body of sportsmen advisory to the commission. The congress had been established after the state legislature, inundated with an average of 225 separate bills each session concerning hunting and fishing regulations, in 1933 authorized the commission to regulate seasons and bag limits and to organize advisory committees. (Like most state legislatures at the time, however, they did not grant authority to limit the sale of licenses or to limit the number of hunters in particular areas.) The game committee of the department, of which Leopold was a member at the time, recommended that a system of hearings be held each winter in each county of the state, at which local sportsmen could discuss the department's proposed regulations and elect a three-man county game committee. The following year the commission made provision for the members of the county committees to attend a state-wide hearing, which in 1938 became known as the conservation congress and acquired an executive council. The purpose of the county hearings and the statewide congress was to provide a forum for democratic expression of local thinking on game matters and to aid in advising the commission on regulations. Over the years the congress also demonstrated a remarkable capacity to weld often divergent local desires into a coherent state game and fish code. This capability was perhaps at no time better demonstrated than during the deer debates of the 1940s. Recommendations of the department and the congress were transmitted to the commission for final decision. The governor retained veto power, but rarely invoked it.[27]

---

tried to dissuade Leopld from accepting the appointment. Leopold's feeling that "a man ought" to do it is corroborated by his son Starker.

27. For background on the congress see Memorandum, W. F. Grimmer to Ralph M. Immell, 13 Dec. 1933, LP 2B10; W. T. Calhoun, "Wisconsin Conservation Congress: Democracy in Wildlife Regulations," WCD Pub. 604 (c. 1941); and Walter E.

There is no doubt that the 1943 conservation congress was engineered toward a vote for an antlerless season. Its chief leaders from the start, Dr. H. O. Schneiders of Wausau and Clarence Searles, a cranberry grower from Wisconsin Rapids, were good friends of Leopold and Swift and went regularly into the deer yards with the Feeney research crew. They understood the importance of scientific game management, perhaps even better than some of the commissioners. They subjected the sportsmen delegates at the congress to a battery of vigorously proreduction spokesmen: R. R. Hill of the U.S. Forest Service told of some $400,000 of damage deer had inflicted on plantations and natural reproduction on the Chequamegon and Nicolet national forests in Wisconsin during the preceding decade; Leopold spoke on irruptions on the Kaibab and in Michigan; Richard Deerwester used population graphs to explain "the Leopold law as I call it, because it is the nearest approach to common sense"; Fred Wilson, chief forester of the conservation department spoke; and the congress saw the movie, *Starvation Stalks the Deer*.

There was earnest opposition on the floor from some of the delegates who favored conservation rather than destruction of the deer, so there would be some left for the soldiers when they got home from war. Winter feeding was the answer, these delegates argued, just as Joyce Larkin had said they would. Indeed, the state legislature had just that year earmarked fifty cents from each one-dollar deer tag for winter feeding and purchase of deer yards, despite the plaint of conservation department officials that winter feeding was a waste of money and only made matters worse. The leaders of the proreduction faction, however, were reinforced from the floor by the Vilas County delegation and a few others, who displayed their new-found trust in the conservation department and reasserted the need for an antlerless kill. This support from Vilas County, following upon Joyce Larkin's "conversion" by Aldo

---

Scott, "Wisconsin Conservation Congress: The First Forty Years," in *40th Anniversary, Wisconsin Conservation Congress* (undated pamphlet, 1974), 1–16. In 1972 the state legislature gave statutory recognition to the congress.

Leopold, was undoubtedly a major factor in persuading other northern delegations to go along with the program for reduction. The counties represented in the congress voted 50 to 17 in favor of an antlerless season.[28]

The department, concurrring in the recommendations of the citizens' deer committee and the conservation congress, recommended a nine-day season on antlerless deer to the commission at its July 13 meeting. Leopold and Moreland, as commissioners-designate, were present at the meeting, but there is no indication of their having spoken. According to the minutes, the commissioners were convinced that the deer herd had to be reduced and felt "that it will be necessary therefore to kill both antlered and antlerless deer in 1943." They directed the department to prepare a new plan for separate consecutive seasons on the two types of deer for consideration at their August meeting. The commissioners' rationale for this seemingly arbitrary directive remains an enigma. Perhaps they felt that a switch from a season on bucks alone to one on antlerless deer alone would be too drastic for the unenlightened public, so they would compromise and have one season of each.[29]

One week before the August meeting of the commission, the state assembly came within one vote, 46 to 47, of accepting a resolution opposing "the proposed slaughtering of deer." As the resolution put it, they were "dissatisfied with, and skeptical of, the findings of the investigating committee." Being "fully aware of the importance of maintaining a maximum number of deer in the state,"

28. Congress transcript, 21 June 1943, WCD files, SHSW Archives. The executive council of the congress recommended a fifteen-day season on antlerless deer, but the congress shortened it to nine.

29. Commission minutes, 13 July 1943. According to the *Sheboygan Press* of 14 July 1943, Schneiders and Searles of the congress were pitted against Dickinsen and Corcoran of the commission at one point in the discussion. Searles was able to extract from Corcoran the admission that he thought bucks ought to be legal as well as does, so hunters would have a chance to get trophy antlers, to which Schneiders retorted, "This is a game management program we're advocating this year, and the trophy hunters can damn well stay at home for one season."

they had provided adequate means for an emergency feeding program and now expected the commission to carry it out.[30]

The department dutifully presented a "split-season" plan (four days of buck followed by four of doe) at the August commission meeting, but with the caveat that they nevertheless favored a strictly antlerless hunt. Likewise the representatives of the congress would not move from their earlier recommendation for an antlerless season. The commission, however, insisted on the split season, a plan favored by no one but themselves. It was their first significant departure from their usual practice of following the recommendations of the congress and the department. The August meeting was to be Leopold's first as a member of the commission, but he was ill and could not attend. The minutes indicate only that Chairman Dickinsen had permission from Leopold to cast his vote in the affirmative. Although Leopold throughout the decade consistently favored herd reduction by means of an antlerless season, as opposed to an any-deer or a split season, he may well have believed that the only alternative to a split season in 1943, given the commissioners' panic about public opinion and legislative reprisal, was a straight buck season with no effective reduction at all. His main concern, in the end, was that the herd be reduced.

The herd *was* reduced. An estimated 128,296 deer, almost three times as large a kill as ever before, were taken in what became known to many as the "crime of '43." Thousands of licenses were sold after the beginning of the season, many to women whose husbands or friends had presumably shot an extra deer. In Vilas County alone, 500 licenses were issued to women, as compared with only twenty-four the previous year. The commission, to be sure, had no authority to stop the sale before the season opened. Heavy snows in some areas forced hunters to concentrate in other places, with the result that certain areas were badly overshot and the deer in others reduced hardly at all. No county was subjected to a heavier onslaught of hunters than Vilas, which had argued most fervently for a con-

30. Jt. Res. No. 104A, 3 Aug. 1943, *Wisconsin Senate and Assembly Joint Resolutions,* 1943.

trolled hunt. The split season and the free-for-all hysteria engendered by wholly inadequate enforcement were blamed for a heavy illegal kill and widespread bootlegging of venison.[31]

An antlerless hunt might have prevented some of the worst abuses of sportsmanship, and a controlled hunt would certainly have better distributed the kill. It would be twenty years before the state legislature would see fit to entrust the commission with authority to conduct a controlled hunt, though it would be requested at practically every session. The commission undoubtedly merited such powers long before 1963, but their seeming capriciousness in setting the 1943 split season undermined their efforts in the legislature for years.

A straight antlerless season would not by any means have averted public revulsion against the slaughter of does and little fawns, particularly after millions of people had just viewed Walt Disney's motion picture classic, *Bambi*, released the year before. An open season on antlerless deer after a long period of protection *was* a slaughter, Leopold pointed out, and sentimentalism would never cease. "Well-intentioned people who think only of deer and not at all about deer range," he explained to a correspondent, "have a psychological urge to protest further killing."[32]

True it is that the Wisconsin public had not been adequately prepared to accept the fact of too many deer. There had been only two articles on deer in the *Conservation Bulletin* in the entire two or three years prior to the 1943 meeting of the congress. A special issue in August containing Leopold's report on deer irruptions, an article by Feeney, and the report of the citizens' deer committee was hardly enough to redress the balance. Leopold, in fact, was profoundly distressed with the August issue of the *Bulletin*, at least with Feeney's article. He had expected a report dealing with the physical facts of the deer herd and

31. This account is drawn from many sources, including Joyce Larkin's letter to Leopold of 31 Dec. 1944, newspaper articles, department reports, commission minutes, and congress transcripts.

32. AL to Wallace Schwass, 6 June 1944, files W. J. P. Aberg.

the deer range, as uncovered by the deer research project, but Feeney instead had provided a commentary on proposed remedies. "The reader has never been told what facts the deer research project has found," Leopold charged. "There was urgent need for such a report as long ago as last winter, but here we are a month after action has been taken [the decision to hold a split season], and the citizen has not yet been told the facts underlying that action." There was, in effect, no functioning machinery for public education.[33]

On the other hand, 1943 was a favorable year for an antlerless season because of widespread starvation the preceding winter and the public attention it had attracted. The citizens' committee and the conservation congress, at least, had been brought to acquiesce in the necessity of herd reduction. To wait for widespread public acceptance would have meant decades, if it ever came, and there *was* a certain amount of urgency.

Even among members of the conservation congress there was a feeling that the commission, by substituting a split season for the antlerless, had sold them out. From an Ashland County delegate, for example:

> You fellows who were there remember in '43 how they built up for us all day long and we had our deer hunting discussion that night. They had these movies of all of the overbrowsing deer, and the starving deer. It was just choked down our throats about the over-population. We had 550,000 deer in Wisconsin, and it had to be cut 50%. We voted then to have an open season on one deer, and you know what happened. They had the split season.[34]

33. AL to E. J. Vanderwall (director of the department), 3 Sept. 1943, copy in files of W. J. P. Aberg. Leopold may have been personally irritated at Feeney's "commentary on remedies," because he felt that that was more properly the function of the citizens' report that appeared in the same issue. But in the letter, he did not blame Feeney: "Good field men often write poor reports, but a good overhead staff does not let poor reports get into print, especially on so crucial a subject as this."

34. Congress transcript, District 2, Rhinelander, 11 June 1945, p. 19, WCD files, SHSW Archives.

When things go wrong, or people think they do, someone or something has to take the blame. After the "crime of '43" it was the split season and "they" (the powers that be) that caught the blame parried by the considerable number of sportsmen in the congress who had in fact voted for herd reduction, albeit by means of an antlerless season. It would be five years before the conservation congress, even under the skillful leadership of Schneiders and Searles and with all the propaganda resources of the department at its disposal, could be brought to recommend another antlerless season.

Reprehensible as the 1943 season may have seemed to the public, the conservation congress, and the legislature, for Leopold and the deer research men it was only a good beginning in desperately needed herd reduction. In an article titled "What Next in Deer Policy?" published in the *Wisconsin Conservation Bulletin* for June 1944, Leopold analyzed the 1943 kill and made prognostications for the future. Using the department's official pre-season estimate of 500,000 deer—a "pure guess" he called it, but argued that the figure was more apt to be low than high—he calculated the 1943 kill as a mere 10% reduction, on the basis of 50,600 does taken. Citing a 90% reduction on the Kaibab, 70% on the George Reserve in Michigan, and 50% in Pennsylvania, he suggested that Wisconsin, "by reason of her prompt start in 1943," might get by on a 60% reduction, i.e., a stabilized carrying capacity of 200,-000 deer. Such a reduction would take five years at the 1943 rate. The penalty of delay in completing this reduction, he warned, was an even lower ultimate carrying capacity (Figure 3).[35]

35. "What Next in Deer Policy?" *WCB*, 9:6 (June 1944), 3–4, 18–19. Leopold prepared a similar calculation (12 per cent reduction, on the basis of a revised doe kill figure of 62,000) for a statement on the excess deer situation nationwide, which was issued by a committee established at the North American Wildlife Conference in April 1944. Walter P. Taylor, chairman of the committee, apparently insisted that the 1943 reduction in Wisconsin be figured as 26 per cent, taking the entire kill of 128,296 (128,296/500,000), rather than does alone—an indication that Taylor may not have subscribed entirely to Leopold's argument

Explaining the situation to delegates at the 1944 meeting of the conservation congress, Leopold emphasized that population estimates could not of themselves be an exact guide to policy. The condition of the browse was the only satisfactory indicator of whether or not an area was stocked appropriately. The purpose of the estimates and the comparisons with other states, he pointed out, was merely to help visualize the problem.[36]

Leopold's calculations were so graphic and his role as chief advocate of herd reduction so conspicuous that by the summer of 1944 he personally was cast as the scapegoat for the "crime of '43":

> The infamous and bloody 1943 deer slaughter was sponsored by one of the commission members, Mr. Aldo Leopold, who admitted in writing that the figures he used were **PURE GUESSWORK**. The commission accepted his report on that basis.
>
> Imagine our fine deer herd shot to pieces by a man who rates himself as a Professor and uses a **GUESS** instead of facts? Mere fawns just out of their spots were sacrificed by our conservation commission. Does, with young already conceived, young, immature bucks, in fact, everything that ran was indiscriminately slaughtered, not by sportsmen, but by a bunch of hungry meat hunters, spurred on by the commission's poison propaganda.

The foregoing is from an open letter to the governor, state senators, and assemblymen from the Save Wisconsin's Deer Committee, published in its official newspaper, *Save Wisconsin's Deer*, Volume 1, Number 1, August 1944. On practically every page of the first issue and in each succeeding issue the arguments for reduction of the deer herd were ridiculed and castigated as the guesses and non sequiturs of an egotistical professor scheming away against the deer from his swivel chair down in Madison. Leopold sub-

---

that only doe removal could affect the rate of increase and thereby reduce a herd. See Leopold's draft in LP 2B5 and the published statement, "Six Points of Deer Policy," *WCB*, 9:11 (Nov. 1944), 10.

36. "Seven Prongs of the Deer Dilemma," typescript, 26 June 1944, 6 pp., LP 6B14.

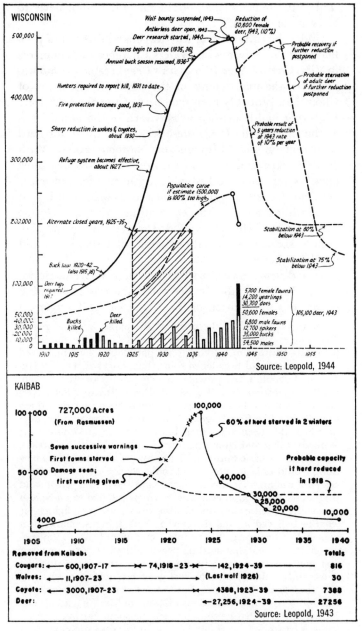

3. *Kaibab and Wisconsin Deer Irruption Histories.*

scribed to the newspaper, but he made it a point never to answer the charges.[37]

Under the editorship of Roy Jorgensen, director of publicity for the Manitowish Waters Chamber of Commerce in Vilas County, *Save Wisconsin's Deer* became the mouthpiece for the malcontents of the north woods—many of whom had probably been members of the Save the Deer Club that had flourished at Hayward in the mid-thirties at the time of the U.S. Forest Service's request for herd reduction on the Chequamegon National Forest. With Joyce Larkin and her *Vilas County News-Review* now openly sympathetic to Leopold's arguments for deer management, Jorgensen probably had a ready-made contingent of erstwhile *News-Review* deer sentimentalists to welcome his new paper. He even provided something for the children—a sixteen-part serial on Bambi of Valhalla, as told by Bambi herself.

Save Wisconsin's Deer stood for conservative conservation and against radical experiments and was dedicated to preventing a repetition of the "crime of '43." Jorgensen

37. In October 1945 the name of the paper was changed to *The Badger Sportsman*. As of 1974 it was still being published (Chilton, Wisconsin), although its philosophy had mellowed and its appeal had broadened somewhat. James B. Hale, a student of Leopold's, has provided a penetrating analysis of Leopold's response to *Save Wisconsin's Deer:* "It's to Aldo's everlasting credit that he didn't fight personalities on the deer question. People like Roy Jorgensen (the editor) were ripe for libel suits had Aldo been so inclined, but he chose to stick to the facts as he saw them and ignore personal insults. This is a reflection, I think, of his basic optimism. I was always convinced he knew things could only improve if he could change the views of the key public figures. He was at heart a kind and gentle person, and tried to see the good in everyone; perhaps you had to know him personally to sense this. But he also was a persuasive fighter for what he believed in, and this is what made him so remarkable a personality, because gentle people aren't often fighters. I'm convinced that he felt that battling personalities like Mr. Jorgensen was much more personally objectionable than accepting the abuse and plodding on. This was good psychology in a way; if you read *Save Wisconsin's Deer* and *Badger Sportsman*, you can find many little indications that one of Mr. Jorgensen's major irritants was that he couldn't get a personal rise out of Aldo and was frustrated as a result. Silence, in this case, did more than a hundred letters by Aldo." Letter to author, 6 April 1970.

and his followers steadfastly maintained that Wisconsin had had only about 200,000 deer, not 500,000, before the 1943 reduction, and that more than three-quarters of the deer in heavily hunted areas such as Vilas County had been killed. Hence their incredulity at Leopold's proposal to continue similar reductions for five more years and their dedication to protecting their deer from edicts of the commission. In fact, they favored abolition of the commission. They also favored strict enforcement of the hunting laws, especially the one-buck law, payment of bounties on wolves and coyotes, and the establishment of large refuges or closed areas (the Vilas contingent favored turning the whole county into a closed area for all game).[38] The entire platform of the Save Wisconsin's Deer Committee was bedecked with flags—they stood, above all, for preservation of game for the boys in the armed services.

The Save Wisconsin's Deer organization, as the name implied, was convinced that there were too few deer. Its members had not even entered the stage of succession to the idea that there could be too many deer. And their position had polarized into a cause that would attract thousands of people.

38. Conservation department officials had in fact recommended a closed season on deer in Vilas County for 1944 because it had been so heavily hunted along the roads in 1943, but the recommendation was not accepted by the commission. Not everyone in Vilas County, however, agreed with Jorgensen's case for a closed season on *all* game, or the manner in which he presented it. This sentiment is made clear in a letter from Alvin Koerner, owner of a resort on Manitowish Waters, to W. J. P. Aberg, 23 Sept. 1944 (Aberg files): "I would not take what you are taking from that D--- F---, which you call a demagog, and that is a nice name for it, for anyone or any price, but I do give you credit that you are willing to back up what you think is right in the face of all of it. He has, however, been proven wrong in some other things in the past, he is now riding on hero worship based on sentimentality by a lot of people, however, I have seen such people come and go, and their fires burn out, so please don't feel as though this whole community is swayed by this 'voice.' I have petitions in my hands right now to back up my request to open hunting except for deer in this area, which will prove to you that the demagog is not 'king' or Dictator."

6

Adventures of a
Conservation Commissioner

Responsibility in a Crisis

Some time in the mid-1940s Leopold penciled the following few lines on the public problem:

'If the public were told how much harm ensues from unwise land-use, it would mend its ways.' This was once my credo, and I think still is a fairly accurate definition of what is called 'conservation education.'

Behind this deceptively simple logic lie three unspoken but important assumptions: (1) that the public is listening, or can be made to listen; (2) that the public responds, or can be made to respond, to fear of harm; (3) that ways can be mended without any important change in the public itself. None of the three assumptions is, in my opinion, valid.

Although the title, "Conservation Education: A Revolution in Philosophy," gives a clue to the direction of Leopold's thinking, the fragment nevertheless ends abruptly on a negative note—one more indication of the quandary in which he found himself when there was no time to wait for a revolution in philosophy.[1]

He explained the predicament to the 1944 conservation congress in his talk, "Seven Prongs of the Deer Dilemma." To those who would favor waiting for favorable public opinion or for better data before chancing another large kill, Leopold said he would reply, "That's no way to talk at a fire":

Deer irruptions are in fact a biological fire, and we who have had our fingers burnt are morally obligated to say so. Irruptions are a slow fire, like those that burned in our

1. "Conservation Education: A Revolution in Philosophy," holograph, n.d., 1 p., LP 6B16.

peat marshes in 1930, and like them, they burn all year and without fuss and feathers, but they do a thorough job.[2]

Leopold's reference to "we who have had our fingers burnt" could apply in a general way to anyone who was aware of the consequences of earlier irruptions in other states. But although he never during the deer debates of the 1940s mentioned Black Canyon, one can read into this statement his consciousness of his personal responsibility for what had happened on the Gila and his burning determination not to let it happen again. Surely his sense of personal moral obligation was a factor in his unwillingness to wait, either for more research or for public opinion.

Leopold's experiences with deer in Wisconsin reaffirmed his conviction as to the responsibility of the ecologist in such situations. "An ethic," he had written back in 1933, "may be regarded as a mode of guidance for meeting ecological situations so new or so intricate, or involving such deferred reactions, that the path of social expediency is not discernible to the average individual."[3] Ecology, he realized, was among the most complex of the sciences and might therefore be the last to achieve the stage of predictable reactions. Yet, committed as he was to deep-digging ecological research, he was equally convinced that the ecologist had a responsibility to "step beyond 'science' in the narrow sense" and offer modes of guidance for meeting ecological problems that were not yet fully understood.

In an unpublished essay on "The Land-health Concept and Conservation," written during the height of the deer crisis, Leopold offered a rueful definition of conservation as "a series of ecological predictions made by laymen because ecologists have failed to offer any." He was pleading for ecological prediction by ecologists, whether or not the time was ripe. "If we wait," he warned, ". . . there will not be enough healthy land left even to define health":

We are, in short, land-doctors forced by circumstances to reverse the logical order of our service to society. No matter

2. "Seven Prongs of the Deer Dilemma," typescript, 26 June 1944, 5 pp., LP 6B14.
3. "The Conservation Ethic," *Journal of Forestry*, 31:6 (Oct. 1933), 635.

how imperfect our present ability, it is likely to contribute
something to social wisdom which would otherwise be
lacking.

What he meant by "prediction," he explained, was "a
shrewd guess" as to the probable conditions necessary for
biotic self-renewal, or health. His own guess as to the conditions necessary for land health included maintenance
of the integrity of the parts, gentleness rather than violence in land use, acceptance of individual obligation
going beyond private profit, and the stabilization of human population density. This was the substance of a land
ethic, as he was later to make clear in his essay by that
title; it embodied the fundamental values, integrity, stability, and beauty.[4]

These values or "modes of guidance" were implicit in
Leopold's analysis of Wisconsin's deer problem in the
1940s, as they had been in his earlier writings, such as the
report on Huron Mountain Club. But whereas the members of the club, sensing that they held lands of unusual
integrity, appreciated his analysis or at least trusted him
and accepted it (and possessed the wherewithal to act upon
it), the Wisconsin public proved unable to comprehend his
message and unwilling to accept it. Because the people
had no conception of what healthy land was, they did not
understand that the state's deer range was sick and the
disease progressive. Feeling keenly his personal responsibility as a citizen, as an ecologist, and ultimately as a conservation commissioner, Leopold pressed nevertheless for
the one thing that to him seemed imperative, immediate
herd reduction, even in advance of public understanding.
In this exigency, as we shall see, he tended to focus more on
symptoms than on processes, to state his case starkly and at
times to overstate, in hopes of spurring people to action.
It is possible, moreover, that his preoccupation with the
public problem in his role as commissioner inhibited the
evolution of his own ecological understanding of what was
in fact a highly dynamic situation. Commissioner Leopold's experiences in the deer debates of the 1940s point

4. "The Land-health Concept and Conservation," holograph,
21 Dec. 1946, 8 pp., LP 6B18.

up not only the necessity but also the difficulty and some of the perils of dealing with an ecological issue in the public arena.

Wolves, Coyotes, and People

The first major issue on which Leopold tangled with the public after the "crime of '43" was the matter of predation. On this issue he at times quite clearly overstated his case, with understandable moral fervor, and was dealt with unmercifully by his critics. By the end of his visit to the Kaibab in 1941, he had turned 180° from his earlier advocacy of eradication to suggest that removal of predators "predisposes" a herd to irruption and to assert that natural predation was "the only precision instrument known to deer management." He clearly recognized predation as but one factor among many and one on which there had been little research, but he did not always so qualify his statements when arguing in favor of using predators to trim the deer population in Wisconsin. When he wrote the report of the citizens' deer committee, he flatly recommended that "a low population of timber wolves be deliberately maintained as insurance against undue congestion or excessive numbers of deer."

Perhaps as an outgrowth of the citizens' committee recommendation, the Wisconsin legislature did not appropriate funds for bounty payments on wolves, coyotes, and foxes in the 1943 budget bill. Violent public outcry after the split season that fall resulted in last-minute passage of a separate bounty bill, but it was vetoed by the governor. *Save Wisconsin's Deer* charged that the bounty bill had been opposed by "the biologists for the Professor," but the commissioners denied having made any recommendation to the governor. In response to the outcry they delegated George Ruegger, a veteran woodsman and trapper, to survey the predator situation that winter. Ruegger reported that there were more coyotes and foxes in the northern counties than he could remember ever having seen; timber wolves and bobcats were also increasing. (Predator populations had undoubtedly been on the upswing before 1943, but the increase had not been noticed.) Ruegger warned that there would be lots of claims for damage to sheep, but he did not believe that damage to deer was anything to worry about.[5] Sportsmen and other citizens in the north country, however, were convinced that the predators were decimating their deer herd, and they deluged the commissioners with pleas and complaints.

It was around this time, in April 1944, that Leopold wrote "Thinking Like a Mountain," though he did not publish it then. Instead, in meetings, correspondence, and articles he repeatedly explained the function of the wolf in his theory of irruptions. He told how the gun was a crude tool for controlling deer populations and how the wolf by comparison was a precision instrument that regulated not only the numbers but also the distribution of deer. He referred to the theory, developed by Paul Errington in his studies of bobwhite quail, that predation tends to be most effective in reducing surplus population to the carrying capacity of the range and is nominal when populations of prey are in balance with their range. Although he admitted that it was not known "whether any such automatic control exists as between deer and wolves," he said it made no practical difference anyway because the wolf population in Wisconsin numbered only a few dozen individuals and could be reduced at will whenever necessary. Behind much of the public criticism of the commission's predator policy was the tacit assumption that the commission's sole function was to furnish more shootable game for hunters. That assumption was fallacious, Leopold pointed out. The commission had the responsibility to harmonize the public interest in forestry, recreation, and game. Deer and cottontails were inflicting severe damage on plantations and natural forest reproduction, and the commission had the obligation to get whatever help it could from wolves, coyotes, and foxes in trimming excess deer and cottontails to mitigate the damage.[6]

5. See *Save Wisconsin's Deer*, Jan. 1945; commission minutes, 14 March 1944; and George Ruegger to Ernest Swift, 7 April 1944, LP 4B4.

6. See especially "The 1944 Game Situation," typescript, 30 March 1944, 6 pp., and related correspondence, LP 6B17. This article was an attempt to express the attitude of the commission on predators. It was not used at the time, but a version of it was

Sportsmen did not appreciate his argument. The Trego Rod and Gun Club in Washburn County, for example, unanimously approved a statement on "The Fox, Wolf and Deer" for submittal to the commission and the press, from which the following paragraphs are exerpted:

> The wolf is the Nazi of the forest. He takes the deer and some small fry. The fox is the sly Jap who takes the choice morsels of game and the song birds. Can Professor Leopold justify their existence because deer meant for human consumption should be fed to the Nazi because we must have that protection for the trees? Can he justify the Jap or Nazi

---

apparently printed in a newspaper, judging from an analysis of similar statements by "the eminent professor" in *Save Wisconsin's Deer*, Feb. 1945. See also "Population Mechanisms of Deer and Quail," Wildlife Ecology 118 course handout, c. 1947, LP 6B15.

There have been relatively few studies since Leopold's day on the effect of wolf predation on deer populations because there have not been enough wolves in most areas to make it a live issue. In general, however, many wildlife ecologists now believe that the quantity and quality of food is the most important determinant of animal numbers, and they would extend this concept to predatory animals also. Thus they would view predators as controlled by the supply of prey, in contrast to Leopold's inference that deer irruptions might be controlled if there were sufficient predators. A large increase in food supply as the result of fire, logging, or other disturbances caused by man might well increase deer numbers beyond the capacity of predators to hold them in check, even though under wilderness conditions wolves might function almost as "precision instruments." There have as yet been no completely documented cases of wolf predation actually controlling deer or other ungulate populations, but there have been a number of studies that suggest beneficial effects of predators on their prey populations. See for example Adolph Murie, *The Wolves of Mount McKinley* (National Park Service Fauna Series No. 5, 1944); L. David Mech, *The Wolves of Isle Royale* (National Park Service Fauna Series 7, 1966); D. H. Pimlott, et al., *The Ecology of the Timber Wolf in Algonquin Provincial Park* (Ontario Dept. of Lands and Forests Res. Rep. 87, 1969); Maurice G. Hornocker, "An Analysis of Mountain Lion Predation Upon Mule Deer and Elk in the Idaho Primitive Area," *Wildlife Monographs*, 21 (March 1970); and L. David Mech and L. D. Frenzel, Jr., eds., *Ecological Studies of the Timber Wolf in Northeastern Minnesota* (U.S. Forest Service Research Paper NC–52, 1971).

because he eats a rabbit or a grouse which are meant for human food, or the song bird on its nest, which was meant by the Lord for our pleasure, because this hungry Jap must live to eat the rabbit to save the tree? Bear in mind that all the time the tree men are not asking for this protection, at least we haven't heard it. . . .

Professor Leopold, can you possible admit that you cannot so regulate the taking of deer by humans, that you must propagate the wolf pack and the fox to eat those deer? It is against the law 360 days of the year to kill deer with a rifle. Such rifle may take deer only some five days a year. It is legal for the wolf and fox to make his kill every day of the year, and they will do this with your blessing, and we understand you are mourning the fact that most states have practically destroyed them. Under these conditions men must go to jail 360 days of the year. We must ask you a question—"Do you like the wolf better than the man?"[7]

Wolves might have remained unbountied in Wisconsin because of the threat of their extermination had the public not confused them with coyotes. Leopold insisted on distinguishing between the two, noting that wolves, although effective at killing deer, were on the verge of extinction, while coyotes were ineffective at killing deer and could not possibly be exterminated. Most men of the north, like Waldo W. Rinehard, an insurance agent from Shawano who exchanged a series of letters with Leopold in March and April 1944, made no such distinction between wolves and coyotes, or dogs for that matter:

About three years ago I went up to the Eberlein Deer Park in Langlade county on Highway 55 to rid that private fenced in area of Wolves. . . . I shot the wolf, and what was it., It was a coyote. A three legged one, with one hind leg missing. Later it was reported that there was another one present. I was not along on that hunt to get the second one. The same group excepting myself shot that one. What Was it. *The neighbors collie dog.*, that had a special place for getting in and out of the fence.

7. J. H. Feil and Harvey Gillette, "The Fox, Wolf and Deer," typescript, c. Jan. 1945, 3 pp., LP 2B10.

Queried by Leopold if he was sure he meant wolves, Rinehard replied:

I am sure that when I talk to you of wolves, I mean just that, and that includes, both wolves, that are present in Wisconsin, *Timber Wolves and Coyotes,* and I do not place the one much further behind or ahead of the other.[8]

In his early statements on the Wisconsin situation, Leopold played down the destructiveness of coyotes. While he had been in the Southwest, as we have seen, he had regarded coyotes as a serious menace to deer, especially fawns, and had urged that coyotes be removed even after he proposed letting up on lions and wolves, but by the 1940s he was pointing out that the presence of coyotes on many irruptive ranges indicated that they were not effective deer predators, otherwise they would have prevented the irruptions. Coyotes were increasing in numbers, he suggested, because they subsisted largely on carrion, of which there had been a superabundance due to winter starvation of deer and abandoned illegal kills. Rinehard, along with countless others, kept after him, however, demanding to know why there were not any fawns if predators were not to blame, and in December 1944, after a smaller-than-expected deer harvest, Leopold wrote to Rinehard that he was now convinced there were too many wolves and coyotes, and he was recommending reenactment of the bounty. Moreover, he was even inclined to admit that he had been wrong in advocating additional hunting of does in 1944. "In short, the predators responded to excess deer more quickly than I thought they had. I'm glad to see

8. Rinehard to AL, 17 March 1944; AL to Rinehard, 24 March 1944; and Rinehard to AL, 26 March 1944; all in LP 2B10. A number of Wisconsin citizens wrote regularly to Leopold, often in a derogatory manner. He always wrote personal replies which were both thoughtful and respectful.

As for the incident about the collie dog, wildlife managers in Wisconsin now regard domestic dogs as far more harmful to deer than wild predators. Rinehard was concerned about dogs too, but as he put it: "If a half fed cur dog will chase deer, what about a hungry coyote, If I were a deer I would rather have cur dogs after me than, Coyotes."

this," he told Rinehard, "because it means our fauna is healthier than I thought."[9]

This was an extraordinary admission for Leopold to make, and there is no satisfactory explanation for it, except that he was making a judgment about field conditions that he had not personally seen, under pressure of intense public outcry about the shortage of deer and the super-abundance of predators. He was in a way a victim of his own too-narrow focus on predators as precision instruments, as well as of the lack of hard scientific data on ever-changing field conditions. For if deer populations seemed to have been trimmed, it was hard to avoid the conclusion that predators had been responsible for it.

On the other hand, Bill Feeney, who was constantly in the field, stated as his opinion that the number of fawns was not much below normal, that there were far too few wolves left to have any noticeable effect on the deer population even in areas where they were most abundant, and that there was no single case where there was any proof that a coyote had killed a fawn. He pleaded that there be no bounty on timber wolves and argued that there was no justification for a bounty on coyotes either, except as a temporary public relations measure. A committee of university professors, which was appointed by the governor to investigate sheep, hog, and cattle losses caused by predation in northern Wisconsin, recommended reinstatement of the bounty on coyotes and foxes to protect livestock interests, but explicitly recommended against any effort either to reduce or exterminate wolves.[10]

Nevertheless, at a meeting with legislators in January 1945, Leopold and the other commissioners agreed to sup-

9. AL to Rinehard, 19 Dec. 1944, files of W. J. P. Aberg. See also AL to Rinehard, 13 Dec. 1944, files of W. J. P. Aberg.

10. W. S. Feeney to AL, 18 Dec. 1944, LP 4B1; and W. A. Rowlands, A. D. Hasler, F. B. Trenk, and J. J. Lacey, "Immediate and Long Term Measures Recommended for Predatory Animal Control in Northern Wisconsin," typescript, 13 Dec. 1944, 8 pp., LP 4B4. Leopold was not a member of the committee and tried not to influence its questions or findings, but he did help the committee get in touch with Stanley P. Young, a nationally prominent expert on predator control.

port reenactment of the bounty on wolves as well as on coyotes and foxes.[11] A bill was passed in March, and in the April *Conservation Bulletin* Leopold explained his new position. He believed a bounty on both wolves and coyotes was now necessary because of the increase in coyote depredations and because it was impracticable for county clerks to distinguish between wolves and coyotes in paying bounties. There was a "probability" that timber wolves had increased, he stated, implying that the species was no longer in immediate danger of extermination. Although he was undoubtedly capitulating to political pressure and public opinion on the bounty issue, Leopold asserted that his aim was still to prevent extinction of the wolf in Wisconsin. "I myself have cooperated in the extermination of the wolf from the greater part of two states, because I then believed it was a benefit," he admitted publicly for perhaps the first and only time. "I do not propose to repeat my error." But in thus salving his conscience he made an unfortunate statement:

> Those who assume that we would be better off without any wolves are assuming more knowledge of how nature works than I can claim to possess.[12]

Under a banner headline, " 'Assume They Know More than I Do'—Leopold," *Save Wisconsin's Deer* quoted the sentence above and commented:

> This statement comes from none other than Professor Aldo Leopold, one of Wisconsin's conservation commis-

11. There was a good deal of legislative pressure on Leopold and the other commissioners around this time, but what effect it had on their stand on the bounty question is unclear. Most of the bills recommended to the 1945 legislature by the commission were killed, in what Leopold and others interpreted as reprisals for the "crime of '43." Leopold claimed, moreover, that he was told in person that a raise in the department director's salary depended on his silence on deer. See his "Adventures of a Conservation Commissioner," typescript, 1 Dec. 1946, 7 pp., LP 6B16, and his "White Tail Deer" (lecture for Game Management 179), typescript, c. 1947, 5 pp., LP 6B15.

12. "Deer, Wolves, Foxes and Pheasants," *WCB*, 10:4 (April 1945), 3–5.

sioners. . . . Read it again because it has that touch of "Leopoldian egotism" and insinuates that he, the great Aldo, places his knowledge above that of any Wisconsin citizen.[13]

Roy Jorgensen, the editor, caught Leopold's every slip and smeared it across the pages of his paper from the first issue, August 1944, until Leopold's death. This was part of the price Leopold had to pay for his unrelenting leadership in the battle against too many deer.

Meanwhile, in the years since Leopold wrote, the timber wolf has been extirpated in Wisconsin, if indeed it had not already effectively been exterminated as he was writing.[14]

Leopold's problems with interpretation and his treatment by the Jorgensens and the Rinehards on the predator issue illustrate the difficulty of dealing with an ecological problem in the public arena, especially when it involves a dynamic situation which one does not himself understand fully. His diagnosis of the problem and his prescriptions for action, however they may have changed during the course of the decade, reflected his continuing search for the conditions of biotic self-renewal, or health. Most men of the north, however, did not appreciate the ecological notion of health; they were convinced that deer were more important than trees and hunters more important than wolves. When there was a fundamental difference in values and an alert, active opposition, the facts of the situation and the interpretation of the facts assumed special importance. But it was precisely the facts and the interpretation

13. *Save Wisconsin's Deer*, June 1945, p. 1.

14. D. Q. Thompson, studying wolves in northern Wisconsin during 1946–1948, reported having observed wolf tracks during seventeen of twenty-five field trips in one county; by 1953, the county forester reported seeing only one track all winter. In 1957 the bounty was discontinued and the wolf given complete protection. As of 1970 the resident population was reported at zero. See D. Q. Thompson, "Travel, Range, and Food Habits of Timber Wolves in Wisconsin," *Journal of Mammalogy*, 33 (1952), 429–42; and S. E. Jorgensen, et al., eds., *Proceedings of a Symposium on Wolf Management*, pp. 4–5. In *Endangered Animals in Wisconsin*, compiled by the Department of Natural Resources Endangered Species Committee, Ruth L. Hine, Chairman (Madison, 1973), the timber wolf is listed as extirpated.

that were subject to change. Wildlife ecology was a new science; its findings would not readily be accepted by the "barbershop biologists" of the north, each of whom considered himself his own expert on deer. In this atmosphere, any reevaluation or inconsistency or "guess" by the scientists lessened their credibility and fueled the self-righteousness of their critics. Waldo Rinehard, in the following excerpts from a letter to commission chairman W. J. P. Aberg, expressed a typical northern attitude toward the "experts":

> One expert [Leopold] says to me in his letter, "From what the field men tell me there are prehaps 12 or 24 timber wolves in the state." If he does not know of his own knowledge whether there are 12 or 12 hundred timber wolves in the state how much of an expert is he.???. . .
>
> As I understand the matter we have experts so that there will be an adequate supply of game, Our experts are interested in one thing apparently, Reduce the Deer, If our experts were willing that we do have an adequate supply of game, then why take off the wolf bounty at a time when it was needed most. . . .
>
> When the experts wake up and find out that there are not so many deer left, and that the wolves have done their dirty work, by that time they will have found, some grey bug that has caused the die off, of deer. The only blessing is the fact that prehaps the experts can teach the deer to spawn so that we can build the herd back in a few years and it will not take 21 years again to do the job. . . .
>
> Prehaps, the big mistake that has been made is the fact that we donot have an open season on experts.[15]

Policy and Public Opinion

Progress in building understanding came slowly in the 1940s. First the "experts" themselves began to comprehend some of the implications of too many deer, then

15. Rinehard to Aberg, 17 Aug. 1944, files of W. J. P. Aberg. The term *barbershop biologist* was used by Gordon MacQuarrie of the *Milwaukee Journal* to distinguish self-styled experts from trained biologists. See "Here Come the Biologists," *Wisconsin Academy Review*, 6:4 (Fall 1959), 157–64.

the conservation commission and the field staff of the department began to understand, and finally some segments of the general public became interested. Leopold as educator was deeply involved in efforts to promote further understanding, but as ecologist–member of the commission he insisted that in the meantime deer policy not be held hostage to public opinion, especially to opinion as registered by vocal interest groups such as the sportsmen in the conservation congress. The commission, he never tired of pointing out, was charged by law with responsibility for harmonizing the general public interest in forestry, recreation, and game. What was the general public interest? For Leopold, there was never any question that the public interest was the stake of present and future generations in a healthy environment. But for the other commissioners, this was an amorphous concept, and as they searched for an embodiment of the public interest they kept coming back to the various vocal individuals and groups who somehow seemed to represent public opinion. The years after the "crime of '43" witnessed a divergence between Leopold and the other commissioners as to the proper response to public opinion, even as they witnessed a convergence of ideas as to the nature of the problem.

In 1943 the deer researchers, the department staff, the conservation congress, and the commission were in agreement on the need for reducing the herd by shooting antlerless deer. But in 1944, as a result of the magnitude of the '43 kill and the turmoil of the split season, the department staff and the congress insisted on a return to the usual forked-horn buck season, contrary to the recommendation of the research crew. As the commissioners were about to vote a motion for a buck season, Leopold challenged their deference to public opinion: "There is an agreement among everybody concerned that overbrowsing is still general and serious and there is no argument for the proposed motion except that the public prefers that particular action."[16] The vote went 4 to 2 with only W. J. P. Aberg joining Leopold in opposition. Again in 1945 the vote was 4 to 2 against Leopold and Aberg.

16. Commission minutes, 6 July 1944.

The other commissioners claimed they were voting for a buck season only because the commission did not have proper authority to conduct an antlerless hunt on a controlled basis. No one understood the necessity of controlled hunting better than Leopold, who had been arguing for it ever since his first publication on "Forestry and Game Conservation" back in 1918. But a quarter century of advocacy in various states had yet to result in passage of any satisfactory bill, and the reason was public opinion. As he was laying the groundwork for a controlled hunting bill to be presented in the 1945 legislative session, Leopold wrote to his former colleague on the citizens' deer committee, Joyce Larkin, to ask whether she thought the northern counties would jeopardize the bill by opposing any grant of additional authority to the commission. His bill entailed authority to limit the number of licenses for specified areas and for deer of specified sex and age classes. "As sure as you live," Larkin warned him, "there will be a roar out of the North which will insist that specifying age and sex means a plot for another antlerless season," and she berated the commission and department for their failure to establish satisfactory public relations. Leopold's reply was characteristic: "I have no illusions about the status of the Commission in public opinion. It seems to me the point is to keep Commission policy wise, fair, and consistent. Whether the public ultimately accepts or rejects the work of the Commission is for the public to decide."[17]

Although Leopold thought the commission ought not to yield to public opinion on the issue of herd reduction, he was not unconcerned about the problem of gaining favorable public opinion. Indeed, he himself wrote article after article for the *Wisconsin Conservation Bulletin*, na-

17. Larkin to AL, 31 Dec. 1944, files of W. E. Scott; AL to Larkin, 8 Jan. 1945, files of W. J. P. Aberg. Larkin felt it might eventually be possible to sell the idea of controlled hunting, one idea at a time, but implored Leopold not to write anything into the bill about specifying age or sex classes. Leopold's proposal as he finally wrote it up ("Controlled Deer Hunting," 5 March 1945, WCD files, SHSW Archives) provided for a regular season and a limited number of permits for individuals or parties to shoot antlerless deer in particular townships in need of re duction.

tional outdoor magazines, and local newspapers; he gave countless talks on the deer problem to everyone from 4-H clubs and freshman engineers to the Society of American Foresters; he painstakingly explained the facts of deer irruptions to hundreds of bewildered or irate, often abusive, correspondents, granting each the courtesy of a direct, personal reply; and he gratefully acknowledged what few letters he received from supporters. He was severely critical, however, of the public relations efforts of the conservation department, especially as reflected in the *Conservation Bulletin*. There was too much of a tendency, he felt, to admit no mistakes, to assume that all conservation problems were soluble, to address articles in the *Bulletin* to the group that made the most noise rather than to individuals who might be induced to practice conservation on their land.[18]

Probably as a result of his criticisms, the commission set up an editorial board for the *Bulletin*, with Leopold a member. They also authorized Ernest Swift, assistant director of the department, to write a popular account of the history and current status of the deer problem, a ninety-six-page bulletin published in March 1946 as *A History of Wisconsin Deer*. Swift, who had been one of Leopold's earliest and strongest allies on the issue of herd reduction, presented a remarkably hard-hitting, incisive analysis of attitudes and policies, as well as the physical status of herd and range, and ended squarely on the public problem: "Mr. Citizen, Mr. Sportsman, the baby is on your doorstep. What do you propose to do with it?" The department distributed its entire stock of 50,000 copies and a half year later reprinted 23,000 more.

At the same time as he was writing the bulletin challenging sportsmen and the general public to come to terms with the deer problem, Swift was overseeing an equally important public relations effort within the department—involvement of the department's own field force in the on-

18. See AL to E. J. Vanderwall, 3 Sept. 1943, AL to W. T. Calhoun, 3 Sept. 1943, W. J. P. Aberg to AL, 4 Sept. 1943, all in files of W. J. P. Aberg; "The Public Relations of the Conservation Department," 18 Oct. 1943, and AL to W. J. P. Aberg, 29 Oct. 1945, both in LP 2B10.

going deer yard surveys of the Pittman–Robertson research group. More than a hundred wardens, rangers, and foresters participated in intensive surveys of 819 known deer yards during the winters of 1944–1945 and 1945–1946. The surveys were evidence that with proper guidance and training the field force could contribute, and they did so with tremendous enthusiasm. It was "the most educational and thought-provoking assignment ever given to our field personnel as a body," Swift stated, adding that if these men had been allowed to assume their proper place in the Pittman–Robertson project from the beginning, most of the antagonism both within and outside of the department would have been eliminated.[19]

Newly converted to the imperative of herd reduction as a result of its deer yard surveys, the department in 1946 finally stepped forward with a proposal for an any-deer season, setting the stage for yet another go-round on the policy and public opinion issue. The conservation congress, even after reconsideration, voted 2 to 1 in favor of the usual buck season, and the commission, fearful of legislative retaliation, sided with the congress against the department. Charles F. Smith of Wausau, an attorney who in the year since his appointment to the commission had become unofficial spokesman for the majority, explained that the deer problem was by no means the whole conservation problem, and the commission was going to need public and legislative support for authority to conduct a controlled hunt and for forestry measures and other aspects of a total conservation program. The gain made in accepting the congress recommendation would more than offset the loss, he said, as the commission could then go to the legislature with "clean hands."[20]

19. "An Analysis of the Deer Problem," submitted by Ernest Swift, in conjunction with H. T. J. Cramer, Allen Hansen, and Ragnar Romnes of the deer committee, typescript, 27 June 1945, 8 pp., files of W. J. P. Aberg.

20. Commission minutes, 9 July 1946, 24 July 1946. Smith's appointment in 1945 had been hailed by *Save Wisconsin's Deer* (March 1945), for which he prepared a statement citing his involvement with the recreational industry in the north and pledging, as part of his general policy, "protection and preservation of our deer as probably the most valuable wild life we

In opposing what he viewed as capitulation to public opinion, Leopold suggested that the commission was not being true to the original role conceived for it when it was established in 1927. "This commission was created, and was given regulatory powers," he reminded his colleagues, "for the express purpose of insulating it from the domination of fluctuating public opinion. It was hoped that such a commission might take the long view, rather than the short view, of conservation problems." What must have cut him to the quick that day was the vote of W. J. P. Aberg, his ally in the movement to establish the commission two decades earlier. Aberg, who had stood with Leopold against the majority in 1944 and 1945, now sided with the others, saying he was unwilling to sacrifice other aspects of the conservation program. The vote was now 5 to 1.

The dilemma confronting the commission, as Leopold defined it in an article for the *Conservation Bulletin*, was whether to reduce immediately, using an any-deer season, an admittedly imperfect tool, or to wait for the legislature to authorize controlled shooting. The last legislature had refused "point-blank," and he was not sure the next legislature would do any differently. There was not much to choose, he observed, "between a recurring series of self-inflicted injuries by the deer, and a recurring series of self-inflicted delays by the representatives of the public will." Aware that citizen groups such as Save Wisconsin's Deer threatened to abolish the commission if it went ahead with herd reduction, Leopold was all for giving it a try anyway, and using it to test the status of public intelligence. "If the public will not tolerate intelligent deer management by its commission," he argued, "then it does not need a commission. The old system of political conservation football would do just as well."[21]

---

have." A few months' association with Leopold on the commission were enough to enlighten him on the biological realities of the deer problem, but Smith had somewhat different priorities and an entirely different conception than Leopold of how to deal with the public. "Frosty" Smith was to become an institution on the commission, serving a quarter century.

21. "The Deer Dilemma," *WCB*, 11:8–9 (Aug.–Sept. 1946), 3–5. Several preliminary drafts of this article were sent to the

Throughout the fall of 1946, Leopold kept trying to get the commission to take a stand on deer policy and confront the public head on. At the root of the public problem he saw two conditions that needed to be corrected, the commission's tendency to keep the public in the dark until the eleventh hour and its failure to ask its critics to take any responsibility for proposing a positive program of their own.

At the September meeting of the commission he introduced a resolution requesting that the conservation congress advise the commission on deer policy and suggesting four possible policy options from which the delegates were to choose. The first provided for continuation of the current buck law, on the stated assumption that there was adequate food for the herd and no significant forest damage. The other three options were all premised on the assumption that there was inadequate food and a consequent necessity for immediate reduction: by controlled shooting (after legislative action), an any-deer season, or an antlerless season. W. J. P. Aberg was apparently opposed to the resolution because he did not want to stir the deer issue at all just then; but the other four commissioners seem, from later correspondence, to have been very much in favor of it. Leopold, however, for some reason unexplained in the minutes, suddenly withdrew it. It may be that Aberg's opposition took him by surprise and disarmed him or that he was seeking unanimity. But it is also at least possible that he withdrew it because he realized suddenly that the commissioners welcomed that form of resolution as a chance to dump the deer problem back in the lap of the

other commissioners for review, since Leopold was attempting to delineate his points of agreement and disagreement with commission policy. C. F. Smith, in a letter of 23 August 1946 (LP 2B10), insisted that Leopold simply state his own views, without ascribing any definite position to the commission. Smith felt he knew the sentiments of citizens in the northern part of the state and they were changing more and more in Leopold's direction, but he did not want in any way to antagonize the north. "They know your position very clearly," he told Leopold. "Our position [as a commission] has never been put to them very clearly. I would rather put it to them myself than have you do it, even though you might do it better than I could."

conservation congress and thus avoid taking a stand on policy as a commission.[22]

The latter interpretation may be supported by his next move. Several days after the meeting, he circularized the commissioners with a "Proposed Deer Policy, 1946–47." Where the earlier resolution was aimed at the congress, the new policy stated flatly, "The Commission will reduce overbrowsed areas in 1947." On the means of reduction the commission was to seek the judgment of interested groups, principally the congress, but it was to insist "that each group must either commit itself to some means of reduction, or else assert that there is no serious over-browsing." As in the earlier resolution, Leopold left no escape for anyone who recognized the problem of over-browsing but was unwilling to commit himself to an immediate, substantial reduction. "It is quite possible, of course, that the Legislature and Congress will reject anything we propose," he granted in his cover letter to the commissioners, "but I cannot sit on the Commission as a professional wildlife manager and let it be said that we have no policy on our most important wildlife problem."[23]

Leopold was a professional wildlife manager. That was a key determinant in his formulation of deer policy, though neither he nor the other commissioners seem to have appreciated its significance or its ironies. No one saw more clearly than he the larger ramifications of the deer problem nor argued more cogently that the deer problem could not be considered alone or the deer hunter would in effect be dominating forest policy. But, although his own thinking about the deer problem was premised on his conception of the general public interest in a healthy environment, the policy options he offered the commission, the congress, and other interested groups were always framed in the simplest terms; they dealt with analysis of the deer food situation and selection of means for reduc-

22. See "Resolution," 13 Sept. 1946, C. F. Smith to AL, 20 Sept. 1946, and V. L. Dickinsen to AL, 27 Sept. 1946, all in LP 2B10.

23. "Proposed Deer Policy, 1946–47," and cover letter AL to V. L. Dickinsen, 17 Sept. 1946, LP 2B10.

ing the herd. Given his professional orientation, his order-
ly mind, and his habit of placing first things first, not to
mention his prior encounters with the practical wisdom
of people like P. S. Lovejoy and Joyce Larkin, his approach
is understandable. Yet on the policy issue he was, in effect,
lending support to the very overconcentration on hunting
regulations and reliance on the congress that he was trying
to overcome. Meanwhile, the more important questions
of *what* was the public interest and *who* represented it
went largely unexamined.

There is considerable irony here when Leopold's formulations of deer policy in 1946 are viewed against his
conception of commission policy two decades earlier. His
proposals in 1946 were remarkably consistent with his
views in 1926, yet the environment of public policy decisionmaking had undergone significant change in the
meantime. Arguing in 1926 for a nonpolitical form of
commission that would be able to provide continuity of
policy in conservation, Leopold had noted a "widespread
misapprehension" of what constituted a conservation policy. A "general public determination" to control fire, for
example, did not constitute a policy on fire. A fire policy
meant choosing one of several possible systems for organizing fire control, and continuity meant staying with the
choice long enough to determine its effectiveness.[24] Policy,
he had in effect been saying, dealt with techniques, or
means, rather than simply with ends. His assumption was
that the public and the commission concurred in the objective. This may have been an obvious assumption in
1926 when the objectives he cited, such as fire control,
game refuges, and public parks, were separate, distinct
goods. But the objective for which he sought acceptance
in 1946, deer herd reduction, was the outgrowth of a new
view of ecological complexity, a view that the general public did not yet share or understand. Herd reduction was
actually a means to a larger end, namely, ecological integrity or environmental health. But framed in its simplest terms the policy of herd reduction seemed to be en-

24. "Organizing Conservation in Wisconsin," typescript, 18
Oct. 1926, 6 pp., SHSW.

tirely negative and therefore a direct challenge to the widespread public desire for a maximum deer herd, conceived as a positive good.

If the public and the commission did not concur in the determination of an objective, what then was the proper role of the commission? Had Leopold framed the question in these terms, he might have formulated policy that looked beyond the immediate goal to broader considerations of the public interest and beyond purely technical means to include such factors as public education, timing, and strategy—or "ecological engineering," in the rubric of P. S. Lovejoy. What was ultimately required, as he understood better than anyone, was a transformation in public attitudes and values, a massive task in education. The department could hardly begin mounting such an effort without broad-based, forward-looking policy guidelines integrating the deer problem and the public problem in a total conservation program. Ernest Swift and a few of his colleagues in the department had begun to realize the need for such guidelines and in a brief "Analysis of the Deer Problem" in 1945 had enumerated a dozen different matters that required integrated policy decisions by the commission. The solution to the deer problem lay in long-range planning, not in the question of what kind of season to have, they pointed out, but there is no evidence that their analysis was considered by the commission.[25]

25. Swift et al., "Analysis of the Deer Problem." Among the matters suggested for policy decisions were the intensity and magnitude of the departmental deer-yard survey, the relationship of deer to forestry and to the resort business, proposals for controlled hunting, the problem of refuges, hunting in agricultural areas, crop damage, winter feeding, fire lanes, acquisition of deer yards, and of course publicity.

The statement explicitly warned that "to point a drive" for any particular type of season in 1946 or 1947 would be the gravest mistake that could be made, since it would immediately defeat past progress and nullify any future findings. This statement may have been directed specifically at one of Leopold's plans submitted two weeks earlier, in which he had urged the commission to adopt a plan, binding on all, "to drive for a one-deer season in 1946" in those areas that needed reduction. See "Proposed Deer Plan, 1945–46," 14 June 1945, LP 4B1.

The irony in Leopold's 1946 efforts to precipitate a deer policy is compounded when one reflects on the kind of men Leopold would have chosen as commissioners back in 1926: "The Commissioners should not be specialists and should not ride personal hobbies. They should be big enough to appreciate all phases of conservation, and to ask for specialized advice when they need it." No one would suggest that Aldo Leopold was not a big man or that he did not appreciate all phases of conservation; but game management, particularly deer management, was his hobby years before it became his profession, and one might suggest that during the decade of the 1940s he was "hung-up" on the incontrovertible fact of too many deer. The deer dilemma for Leopold was twofold. It was the dilemma of a scientist who had left his academic chair to take a stand in the public arena, and the dilemma of a policymaker who felt compelled to formulate policy on purely technical, scientific grounds.

Scholars of resource policy and conservation history have noted the remarkable extent to which scientific research, technical data, and criteria of efficiency have been the basis for natural resource policy decisions in the United States. In large part this is owing to the influence of Gifford Pinchot, W G McGee, and other leaders of the early conservation movement, including Aldo Leopold himself. This reliance on scientists and experts has imparted a strongly rational and professional cast to policies and programs. But it has not been entirely salutary, for the requirements of specialized methodologies have often encouraged compartmentalization of decisions, overconfidence in the correctness of policies so derived, and a neglect of "nonscientific" or nonquantifiable elements. "There is reason to believe that in no other area of public policy has the practice of relying on research and scientific data as a basis for decision been more pronounced," writes Norman Wengert in his classic short study, *Natural Resources and the Political Struggle.* "In no other field, however, is there greater confusion between the role of the scientist in providing information relevant to policy decisions and the role of the political process in combin-

ing such information with value judgments, program
goals, and preferences in order to reach a policy decision."[26]

There were other "big" men on the Wisconsin Conservation Commission in the 1940s, as many as at any other
time in its forty-year history, and they were not specialists
in resource management. There is no question that these
men, businessmen and lawyers, respected Leopold for his
knowledge, experience, and analytical skills and that they
listened sympathetically to him and learned a great deal
from him, including the effects of overbrowsing, the fu-
tility of winter feeding, and the desirability of herd reduction. In private conversation they doubtless agreed
with him point after point, thus implicitly encouraging
him to seek new bases of accord, yet when it came to official meetings of the commission time and again they left
him standing alone. Their instincts as lawyers or businessmen told them they could not simply ignore public
opinion in following the lead of Leopold and department
specialists; yet, not one of them accepted Leopold's challenge or the request of Ernest Swift and other department
officials to formulate an alternative policy. Much as they
may have insisted that the deer problem was exaggerated,
it did in fact mushroom all out of proportion to the rest
of the conservation program because the commissioners,
hoping it would simply go away, failed to come to terms
with it.

Leopold made several more attempts that fall to formulate an acceptable policy, but the commissioners did not
adopt a policy on deer. They did not even introduce a controlled hunting bill in the next legislature. And they did
not, for all their subservience, secure passage of their top-
priority forestry bill to regulate cutting of immature
stands. Instead they were treated to another legislative
joint resolution, No. 54,S. "The people of Wisconsin,"
one clause stated, "are gravely concerned" about starvation losses and "are further dissatisfied with, and skeptical of, the findings and competency of testimony offered

26. (New York, 1955), 3–4. See also Samuel P. Hays, *Conservation and the Gospel of Efficiency: The Progressive Conservation
Movement 1890–1920* (Cambridge, Mass., 1959).

by proponents of the one-deer [either sex] proposals."
The legislature, mindful that the recreational business
was the state's second-largest industry and was "jeopar-
dized by the proposed general slaughter of its deer herd,"
reminded the commission of its responsibility to operate
an adequate winter feeding program and admonished it
not to forsake the buck law lightly:

> This legislature recommends that the conservation com-
> mission adhere to and reaffirm the traditional and success-
> ful policy and law of this state governing the killing of
> mature male deer unless any order by the conservation com-
> mission authorizing the killing of deer of either sex is first
> approved by the county board of any county affected by
> such order, before such order becomes effective in such
> county.[27]

The 1947 conservation congress voted a buck season,
and the department, as in 1946, recommended an any-
deer season. Ernest Swift and ten of his colleagues in the
department who were most involved with the deer issue
prepared another statement to the commission. Fearing
that the commission might once again favor the congress
recommendation over the department's, Swift's group
wanted particularly to emphasize the limitations of the
congress as an advisor to the commission on the issue of
deer reduction. For one thing, as Leopold had pointed
out repeatedly, the congress could not possibly be considered as representing all the interests at stake. The burden of their argument, however, was that in an emergency
situation, where the threat is of a technical nature not
easily understood by laymen, it may be imperative to act
immediately on the advice of technical men and make
explanations to the public after the fact. In 1945 when
Swift and his colleagues called for a broad-gauged, integrated approach to the deer management–public relations
problem, the department was still recommending the usual
buck season and clearly was not crisis minded. By 1947
they, like Leopold earlier, had become convinced that a
crisis situation existed and were now willing to disregard

27. Jt. Res. No. 54,S, 23 May 1947, in *Wisconsin Senate and
Assembly Joint Resolutions, 1947.*

the congress and public opinion and opt for immediate reduction.[28]

The commission, as in 1946, voted 5 to 1 for a buck season as recommended by the congress. According to the minutes:

> Commissioner Dickinsen [the chairman] stated that the commission had created the congress and he did not think

---

28. "The Wisconsin Deer Problem as of 1947," typescript, n.d., 3 pp., WCD files, SHSW Archives. This statement bears the unmistakable imprint of Aldo Leopold's thinking on the deer problem at that time. It may have been written by Irven O. Buss, a former student of Leopold's, who was appointed chief of wildlife research in the department after his return from the war. By 1946–1947, a number of Leopold's students, back from the war, were working on the deer project and in other phases of a growing conservation department research program, and their voices undoubtedly lent support to sentiment within the department for herd reduction.

To illustrate the role of technical men in a crisis situation, the report cites the preventative killing of cattle by the U.S. Department of Agriculture in 1922 and again in 1947 without first consulting cattlemen, because of the imperative need to prevent the spread of hoof-and-mouth disease. Leopold had used another facet of the same incident, the slaughter in 1924 of 22,000 deer in a herd inflicted with the disease in the Stanislaus National Forest in California, to illustrate the possibilities of technical management in "Southwestern Game Fields" (1927, Ch. I, pp. 31–32, LP 6B10). "Of course the primary incentive was protection of cattle," he had written, "but it shows that when we really want game management badly enough we can get it."

It was an example, unfortunately, that could cut both ways. During 1940–1944 Leopold found himself on the other side in a similar case in Florida, where the Bureau of Animal Industry decided to kill all the deer in large areas of the state where they were allegedly carrying a fever tick dangerous to cattle. The killing program was finally challenged by the Commissioner of Indian Affairs when it reached the Big Cypress Seminole Indian Reservation, where Indians had been depending on the deer for centuries. As the National Audubon Society's representative in the controversy, Leopold wrote a preface, "Ticks and Deer: A Lesson in Conservation," for a never-published report challenging the technical management program of the BAI. "This episode," he wrote, "shows that a scientific bureau, confronted with a question of wildlife eradication, may prefer to lose the wildlife than to lose time in scientific research for alternatives" (LP 6B16).

it fair for its members to come down at their own expense, make recommendations in all sincerity, and then have the commission vote against their recommendations.[29]

What could one say to an argument like that?

Defining the Public Interest

The public interest is no simple, static entity. It is a function of community, and one of its purposes is to promote effective interlock among the various elements of the community, all of which are in flux. As an expression of the political process, therefore, the search for the public interest depends to a great extent on the perceptions, values, and interrelationships of the various individuals and groups that comprise the decisionmaking environment of a given issue at a given time.

Aldo Leopold's conception of the public interest was much broader than his attempted formulations of deer policy might indicate. Indeed, his insistence on herd reduction was motivated not so much by concern about deer as by concern for the forest and related environmental values that he believed were not being articulated by the various vocal interests in the controversy. His view of the general public interest was based on a conception of community broad enough to encompass elements of the natural environment as well as human society, with a time horizon spanning centuries rather than years. He believed the conservation commission had a responsibility to identify and weigh these values of more general interest along with those brought to its attention by special interests. But in the decisionmaking environment of Wisconsin in the 1940s his views found scant support.

The public interest in a healthy environment in northern Wisconsin, Leopold believed, entailed an effort to restore a naturally reproducing conifer–hardwood forest with its full complement of native flora and fauna. To the conservation congress, the commission, and anyone else who would listen, he explained that the test of adequate

29. Commission minutes, 1 July 1947, p. 23. Virgil Dickinsen was an insurance agent from Augusta in northwestern Wisconsin.

reduction in deer numbers in Wisconsin would be the ability of pine seedlings to survive the deer. Pine—white pine—because that was the characteristic tree in northern Wisconsin in presettlement times. Leopold talked hemlock too, and yew, yellow birch, white cedar, red maple, all of which were being run out by deer, together with a host of other flora and fauna. But first prize on twenty million acres of Wisconsin conifer–hardwood forest that had greeted the lumber barons in the second half of the nineteenth century was white pine, an estimated 130 billion board feet of it, less than 1 per cent of which still stood by the time Leopold arrived in the state in 1924. Surely there was an obligation to encourage reproduction of substantial acreages of such forests. Was that not a major responsibility of the conservation commission, the reason for the hundreds of thousands of tax dollars entrusted to it for the administration of state forests, fire control, and the forest crop law? Viewed in economic terms, the principal all-public interest for which the commission was responsible in the 1940s was forestry, supported by a two-tenths mill tax on property. Leopold did not miss the policy implications of this financial arrangement, even though he failed to write them into his statements of proposed deer policy. If deer were allowed to destroy natural reproduction and new forest plantations and to change succession to commercially inferior species, the commission would be responsible to the public for the loss.[30]

30. For insightful analysis of public policy implications in Wisconsin's nineteenth-century experience with the lumber industry see James Willard Hurst's monumental *Law and Economic Growth: The Legal History of the Lumber Industry in Wisconsin, 1836–1915* (Cambridge, Mass., 1964).

Leopold pointed to some of the public policy issues following an examination in May 1944 of deer range in the Saddle Mound area of Jackson County in central Wisconsin. The area had been planted to pine by the CCC in the 1930s and Leopold had visited it himself around 1934, at a time when he was concerned that game habitat requirements were not receiving enough consideration in department and CCC forestry programs. Now in 1944 he reported that because of severe overbrowsing by deer, both plantings and natural reproduction of all commercial species, especially jack and Norway pine, were "dead, dying, or damaged." Since this was county forest land subsidized under

Not only were there moral and economic reasons for securing reproduction of white pine, but there were also ecological pressures for doing so immediately. In order to get a "catch" (large-scale natural reproduction) of white pine, Leopold explained, five conditions had to be met simultaneously: a "seed year," which was largely dependent on weather conditions and occurred only once or twice in a decade, fire protection, the proper amount of shade, mineral soil, and a low in the snowshoe hare cycle. Let excess deer complicate the situation and the odds against the pine were virtually prohibitive. If Wisconsin failed to get a catch of pine during the decade of the 1940s, he asserted, it would have to wait half a century for another chance. The even-aged aspen, birch, and other hardwoods that had seeded in after the advent of effective fire protection in the early 1930s provided proper shade conditions for a catch of white pine in the mid-forties, but they were fast closing their crowns, and then there would not be another chance for large-scale natural reproduction of pine until after the hardwoods were cut. Meanwhile, the deer were busily chewing up pine plantations and whatever natural seedlings they could find. In voting on the deer question the commission was making a decision, Leopold stated, as to whether Wisconsin was going to have pine reproduction in this century.[31]

---

the Forest Crop Law it was obvious "that both county and state are piling up a growing liability on a shrinking asset, not to mention the wastage of the CCC investment." He noted much talk of "letting the deer have this region, i.e., giving up hope of forestry," but he strongly disapproved. For one thing, the area's legal status and economic structure would have to be changed. For another, the present increase in deer numbers and resulting destruction of browse species would doom it even as a straight game area. Most crucial, letting the deer have the area would mean "recognizing the segregation of game lands from forest lands," when public forest areas were small to begin with and the entire structure of the conservation department was premised on multiple use. Leopold's analysis of policy issues before the commission in 1944 was entirely consistent with his analysis a decade earlier. ("Deer Range of Saddle Mound Area, Jackson County," typescript, 4 May 1944, 4 pp., LP 3B1. See also commission minutes, 6 July 1944, p. 10.)

31. Commission minutes, 9 July 1946. Leopold tended to

The usual argument from north country citizens, and even from some department field men, was that the forest was growing up despite the deer. Said Leopold, "I would ask: forest of what? Are white birch and popple a forest?" His conception of a forest was qualitative, and he instinctively recoiled at the thought of vast acreages of white birch and aspen, which he regarded as weed species, of little commercial value and unpalatable for deer.[32]

When his arguments failed to persuade the other commissioners that the public interest in forests outweighed the loudly trumpeted interest in deer and required immediate reduction of the herd, Leopold attributed his difficulty to the fact that the forest interest was "silent." There was no public clamor, no organized effort to preserve the forests from the deer as there was to preserve the deer themselves—except of course for the U.S. Forest Service, which never gave up its campaign for herd reduction on the two national forests in Wisconsin. Leopold was disturbed that so much attention was paid by the commission and the public during the 1940s to representatives of the Wisconsin Cranberry Growers Association, who claimed perhaps $20,000 in deer damage a year, when no one worried about the future forests, which were suffering ten or a hundred times that much damage each year from overbrowsing. "The difference," he suggested, "is that there is no 'Wisconsin Pine Growers Association' to call attention to depredations by surplus deer. The future forest belongs to everybody, and hence to nobody."[33]

The commercial forestry interest was "silent" in the

dramatize the odds against the pine, especially the role of deer in inhibiting successful reproduction. There was another factor that he did not mention—white pine blister rust—the fear of which has discouraged many foresters from even thinking about restoring white pine in the Lake States. The blister rust problem, however, is not insurmountable. See Ralph L. Anderson, "A Summary of White Pine Blister Rust Research in the Lake States," USDA Forest Service General Technical Report NC–6, 1973.

32. "Seven Prongs of the Deer Dilemma," speech at conservation congress, 26 July 1944, LP 6B14. For a discussion of changes in the evaluation and uses of aspen see pp. 254–55.

33. "Mortgaging the Future Deer Herd," *WCB*, 12:9 (Sept. 1947), 3.

1940s because it was neither economically impaired by overbrowsing at the time nor conscious of damage to its long-term best interests. The deer did not eat the wood. After the white pine was gone the big lumber companies, far from worrying about growing new trees that would require a hundred years or more to reach sawlog size, moved on to virgin forests farther west, while in northern Wisconsin, gutted and depressed, interest shifted to lower-quality and second-growth trees that could be cut for pulp. Such logging as occurred still followed the exploitative pattern of cut-out and get-out, but by the 1940s some 70 per cent of the total acreage cut annually was immature second growth. The struggle in the forest industries in the 1940s was between hundreds of small timber operators whose very livelihood depended on their continuing to cut, whether the available stands were mature or not, and the huge Wisconsin and Fox valley paper mills, which were already going outside the state for nearly 80 per cent of their pulpwood (to satisfy the rising wartime demand for paper products) and were concerned about the future supply. The mills, looking ahead for maybe five to ten years, advocated legislation to regulate the cutting of immature stands, on private as well as public lands, so that second growth would not be cut during the years when it would be laying on its greatest board-foot increment. Neither the pulp cutters nor the paper companies worried about seedlings, certainly not white pine seedlings, and hence they did not worry about deer.

Leopold's colleagues on the conservation commission were men of affairs, and they viewed the public interest in forestry in more immediate, more political terms than he. Concerned about the economic health of the north country, they characteristically tended to support the big paper mills, the major employers in the area, and hence regarded legislation to regulate the cutting of immature stands as more important by far than immediate reduction of the deer herd. While Leopold argued that by failing to provide for herd reduction the commission was sealing the doom of the future forest, the other commissioners countered that by voting for herd reduction when everybody (especially the state legislature) was opposed to it, they would be jeopardizing passage of their bill to control cutting of immature trees in the present forest. "Why spend your state tax money to keep forests from burning," C. F. Smith asked citizens at a special commission meeting on forests in December 1946, "and then see them cut down before they are mature, before you have any seed, before you can see any natural reproduction?" The irony of this argument, as Leopold must have appreciated if Smith did not, was that new natural reproduction was doomed anyway because of excess deer. Smith, from Wausau in the heart of the paper-mill country, had been unwilling to put the commission on the line on the deer issue, but on the forest issue he echoed Leopold, saying, "Get another conservation commission if you will. Get a commission in which you have confidence. But do not throw an idea of this kind out of the window simply because you may not have confidence in some men on this particular commission."[34]

The real threat of deer to future forests was finally demonstrated conclusively by the conservation department during an unprecedented survey of deer damage to forest reproduction in 1947–1948. Set up at the instigation of Leopold, Ernest Swift, and a number of department foresters and game men, with help from one of the top forest statisticians in the nation, the survey applied mechanical sampling methods to some 3,000 miles of survey courses, 11,000 sample plots, and nearly 160,000 individual seedlings of commercial tree species in northern and central Wisconsin. This was a statistical sample of the entire area of managed forest lands, not just an estimate of conditions in yarding areas, as earlier surveys had been.

The results were dramatic and irrefutable. Of seedlings between one and eight feet in height, which had managed to survive all the other vicissitudes of site, weather, birds, and animals, approximately half were heavily damaged by deer, with the result that reproduction of practically all commercial species was now well below the minimum desirable stocking per acre. By comparison, on the Bad River, Lac du Flambeau, and Menominee Indian reservations,

34. Commission minutes, 9 Dec. 1946.

where Indians kept the deer population trimmed to low levels by year-round hunting of both sexes, less than 10 per cent of the seedlings showed signs of heavy browsing. Fewer than one out of twenty cedar seedlings were damaged on reservations, while seven out of ten were damaged elsewhere. No hemlock seedlings above a few inches in height could even be found outside the reservations. Fire was each year damaging one of every 500 acres of forest in Wisconsin; deer were damaging the equivalent of one in five. Forest nurseries were producing several million seedlings a year; deer were annually crippling over 660 million. Such statistics, graphically presented species by species and area by area, would begin to make foresters within the conservation department and in an increasingly professionalized industry more conscious of certain economic aspects of the forest interest in the deer question, though many of the ecological and esthetic considerations posed by Leopold have yet to be heeded.[35]

If the forest interest was "silent" during most of the 1940s, the recreation interest was not. Northern Wisconsin resort owners and their clientele, including local newspapermen, chambers of commerce, and county boards, as well as summer vacationists, were among the most effective agents in arousing public opinion against the slaughter of does and fawns and creating widespread resentment in the north against the conservation department. The recreation interest, represented most conspicuously by Save Wisconsin's Deer, was too vocal, too intransigent to win much support from the conservation department. If it had been so inclined, the department could undoubtedly have cooperated more than it did in habitat management, to attract deer to certain known areas where they would be readily visible to tourists. Leopold had pointed the way

35. See Stanley G. De Boer, "The Deer Damage to Forest Reproduction Survey," *WCB*, 12:10 (Oct. 1947), 3–23; and Ernest Swift, "Wisconsin's Deer Damage to Forest Reproduction Survey—Final Report," WCD Pub. 347 (1948). In 1948 the commission established a Forest Advisory Committee composed of leaders in the state's forest industries, which by the early 1950s became a vehicle for expression of the commercial forestry interest in the deer problem.

in his 1934 meeting with the Chequamegon foresters and in his Rockford and Huron Mountain reports. Similar programs of food plantings in roadside clearings and on fire lanes would become a major part of the department's habitat management effort in the late fifties and sixties, but the attitude of some officials of the department in the 1940s toward the general idea of catering to the tourist interest was betrayed by H. T. J. Cramer, chairman of the department's deer committee, when he queried:

> Do we want to break this free wild spirit by placing the animal on the dole and feeding it like we feed our animal slaves, the horses, the oxen, the cows and what have you, making a semi-domesticated animal out of it, so that smug, comfortable people used to zoos and museums may look at it from the soft seats of their cars without going to the exertions which nature usually demands in return for a look at its finest handiworks?[36]

Although he would probably have favored a program of plantings, if coupled with over-all reduction of the herd, Leopold was never very friendly toward the recreation business, which he considered simply a private commercial interest. He was particularly incensed at the notion, abroad since the 1930s, that recreational advertising was a proper function of the conservation department—especially when the department as yet did so little to enhance the resources it was promoting. A bill introduced in the 1947 legislature would have diverted $100,000 in license funds for advertising and another $230,000 or more for bounty payments, which were also desired by the recreation interests. "This is a plain racket," Leopold protested in a letter to his friend Tom Coleman, chairman of the state Republican party:

> I realize of course that good men differ in their views on some of these questions. I allow for that, but I have been watching the collective performance of the majority party

36. H. T. J. Cramer, "Harvest of Deer in Wisconsin," *Transactions*, 13th North American Wildlife Conference (8–10 March 1948).

in all conservation matters, and I must say that I never saw
a performance so utterly devoid of any guiding principle,
whether my principle or somebody else's. My confidence is
severely shaken, and if this bill passes, it will be gone for
good.[37]

The bill did not pass, but it was the same legislature that
admonished the commission to adhere to "the traditional
and successful policy of killing mature male deer," on the
grounds that the "proposed general slaughter" jeopar-
dized the recreational business. It was in the context of
this credence paid by the legislature to the demands of the
recreation industry that Leopold asserted the primacy of
the public interest in forests and the other commissioners
took refuge in the advice of sportsmen in the conservation
congress.

The other commissioners may, in fact, have believed
that the conservation congress satisfied their quest for an
identifiable public, representative of the public interest.
The 1940s were a time when generalized conservation was
not a matter of prime public concern and statewide conservation organizations did not flourish in Wisconsin. The
Wisconsin Division of the Izaak Walton League, through
which Leopold and Aberg had promoted the Conservation
Act of 1927, was debilitated by the depression; World War
II gutted the Wisconsin Conservation League, a federa-
tion established in 1940. In the partial vacuum created by
the demise of effective statewide organizations and the
war-induced dissipation of general public interest in conservation, narrowly conceived protest groups such as Save
Wisconsin's Deer became even more influential. The public relations division of the conservation department, itself pared to a minimum during the war, could not begin
to deal with the protest element, much less with the dis-
engaged majority.

The conservation congress was clearly the one best hope
of the commission and the department during the 1940s.
It provided already established channels of communication with the sporting public who, after all, constituted

37. AL to Thomas E. Coleman, 3 June 1947, LP 2B10.

a large segment of the department's constituency. Although the commission was charged with representing the general public interest, the department drew financial support for its game management program from sales of hunting and fishing licenses and quite naturally identified with its clientele. It is significant that Aldo Leopold as early as his *Game Survey of the North Central States* (1931) had argued that the average citizen as well as the hunter has a stake in wildlife and that there should thus be "a logical division of conservation liabilities between sportsmen and the general public." Until funds were provided from general taxation for betterments serving the public interest in wildlife, however, the commission could not be too severely censured for identifying with the sportsmen in the congress.[38]

It could be argued that the sportsmen did not represent their own best interest—that their own interest, clearly perceived, would have coincided more closely with the general public interest. Sportsmen were the original conservationists. They believed that adhering to the time-honored one-buck law would maintain a maximum deer herd and satisfactory hunting on into the future, whereas killing does meant reducing the herd and hence the take in future years. But in fact, they not only lost to starvation animals they could have shot during the hunting season, but through their license fees they also bore the burden of winter feeding and deer damage claims, which mounted steadily throughout the decade—all to protect the deer so they could further diminsh the future carrying capacity of the range for deer.[39]

Such was the environment of public interest decision-making in Wisconsin in the 1940s.

38. For an excellent presentation of the rationale for more general sources of funding as related to a broadened conception of the public interest see A. Starker Leopold, Irving K. Fox, and Charles H. Callison, "Missouri Conservation Program: An Appraisal and Some Suggestions," An Independent Study Funded by the Edward K. Love Conservation Foundation, St. Louis, 1970.

39. See W. E. Scott, "Administrator's Dilemma—Sportsmen's Burden," *Michigan Conservation*, 17:11 (Nov. 1948), 6–7, 12–13.

Wisconsin was not the only state with too many deer in the 1940s. The problem of "irruptions," which had come dramatically to attention in the 1920s in certain areas of the Southwest, in Pennsylvania, and in Michigan, plagued many areas of the nation by the 1940s. At national wildlife meetings, Aldo Leopold compared notes with deer researchers and wildlife administrators from other states and took the lead in identifying common denominators in the irruption experiences of the various states and points of agreement on policy.[40] From these contacts with colleagues experiencing similar problems with deer and people in other states he drew strength to continue his struggle in Wisconsin and conviction as to the importance and rightness of the cause. Yet his emphasis on the phenomenon of irruptions, based on his familiarity with the experiences of other states, may have obscured his perception and analysis of ecological processes and environmental changes in his own state of Wisconsin.

As a result of interest and concern about deer problems demonstrated by more than a hundred people at an impromptu meeting at the North American Wildlife Conference in 1946, Leopold decided to prepare for publication a map and analysis of deer irruption areas throughout the nation that he and several of his students had been working on for use in the classroom. The resulting "Survey of Over-Populated Deer Ranges in the United States," based on literature and correspondence with every state in the nation, revealed that of forty-seven states with deer, thirty registered deer troubles. The only region without irruptions was the Southeast. "Here," Leopold wryly observed, "screw worm and hound dog seem to perform the regulatory function elsewhere delegated, often without success, to legislatures or conservation commissions." In addition to mapping and giving nutshell histories of problem areas, Leopold again, as in his 1943

40. See "Six Points of Deer Policy," *WCB*, 9:11 (Nov. 1944), 10. Similar statements were published in the conservation department bulletins of other states.

report for the Wisconsin Academy, analyzed the phenomenon of irruptions. "This paper," he wrote, "proceeds on certain assumptions, which collectively may be called a theory of irruptions. We know of no way in which these assumptions may be proved or disproved at this time."[41]

He assumed, first, that deer irruptions were a distinctly different phenomenon from ordinary winter die-offs. Whereas the latter were governed wholly by weather, could occur at low population levels as well as high, and did only temporary damage to the range, irruptions seemed to occur only at high population levels and to be "cumulative in both timing and range damage," somewhat irrespective of weather conditions. He assumed, further, that the sequence of events first recognized and described on the Kaibab, and later repeated elsewhere, represented "the 'normal' sequence of events in an irrupting herd." The term *irruption*, he granted, was better applied to lemmings, grouse, quail, and other species that exhibited excessive density followed by mass movement and no return of the movers. Deer, he noted by contrast, "are conspicuous for their lack of movement despite excessive density. . . . We use the term for deer because we know of no better one."

The use of the term *irruption* was somewhat peculiar to Leopold. He had defined it in *Game Management* (1933) as a type of population curve exhibiting "severe but irregular fluctuations of no fixed length or amplitude" and occurring but seldom. He did not ordinarily apply the term to populations of deer until his trip west in 1941, but thereafter he used it continually. Other writers employed the term occasionally during the 1940s, perhaps deriving it from Leopold's various publications, but in the years since his death it seems gradually to have faded from use in deer literature. D. I. Rasmussen, who had described the Kaibab sequence during the 1930s without reference to an irruption, suggested that Leopold,

41. AL, Lyle K. Sowls, and David L. Spencer, "A Survey of Over-Populated Deer Ranges in the United States," *Journal of Wildlife Management*, 11:2 (April 1947), 162–77. Popular versions of the technical paper were carried by national outdoor magazines.

with his flair for use of words to provide vivid descriptions, may have selected the word in order to call increased attention to serious overpopulation problems and also perhaps to tie in with current ideas concerning solar or other externally triggered cycles. Semantically, use of the term *irruption* was comparable to later emphasis on the human *population explosion.*[42]

Aldo Leopold's emphasis on the phenomenon of irruptions in his analysis of deer ecology in the 1940s was probably related to his efforts to promote a policy on reduction of the deer herd in Wisconsin during the same years. They were as diagnosis and prescription and, as such, rather uncharacteristic for one so committed as Leopold to a science of land health and so impatient with mere doctoring. The fundamental research on deer productivity and life history under normal conditions, which he had been trying to encourage in the Gila in the 1920s, in Wisconsin in the early thirties, and then in the Chihuahua sierra, had yet to be begun. But the disease of irruption, with its own "normal" pattern of symptoms, had come on and was progressive. Although he may not have appreciated the extent to which his own analysis of the deer situation was affected by his focus on irruptions, Leopold was aware of the basic predicament. In an essay arguing for wilderness as a base datum of normality, he had written:

> In general, the trend of the evidence indicates that in land, just as in the human body, the symptoms may lie in one organ and the cause in another. The practices we now call conservation are, to a large extent, local alleviations of biotic pain. They are necessary, but they must not be confused with cures. The art of land-doctoring is being practiced with vigor, but the science of land-health is a job for the future.[43]

42. D. I. Rasmussen, letter to author, 31 March 1970. For a recent discussion of the phenomenon in an article which includes a critical analysis of Leopold's interpretation of the Kaibab irruption see Graeme Caughley, "Eruption of Ungulate Populations, With Emphasis on Himalayan Thar in New Zealand," *Ecology*, 51:1 (Winter 1970), 53–72.

43. "Wilderness as a Land Laboratory," *The Living Wilderness*, no. 6 (July 1941), 3.

When Leopold's efforts to precipitate a policy on herd reduction in the 1940s are viewed in the context of his life work and of events subsequent to his death, one of the remarkable features is surely his avoidance during that decade of any consideration of the possibilities of environmental management for deer. By the early thirties, he was already concerned almost wholly with the environmental stage in the sequence of ideas concerning game management. His Rockford and Huron Mountain reports, written after his mid-decade shift in intellectual emphasis, demonstrated the validity of his contention years earlier that the techniques of environmental management, once developed, could be applied as readily to problems of too many as of too few. Throughout the thirties he had advocated dispersion of hardwood and coniferous forest types, maintenance or creation of openings for sun-loving browse species, experiments on regeneration of cedar in yarding areas, and consideration for the needs of deer in timing and spacing of logging activities. Many of these same measures, refined somewhat over the years by a modicum of research and demonstration, would be encouraged in the Wisconsin Conservation Commission's deer management policy three decades later, in the 1960s. By then, much winter food of the herd would be supplied by current cuttings. But during the decade of the forties, in the heat of the battle for reduction of Wisconsin's irruptive herd, there is no evidence that Aldo Leopold ever talked publicly about environmental management for deer, even as a hope for the future.

Leopold's emphasis on the ecological perils of too many deer and his lack of attention to the ultimate desirability of environmental management are attributable at least in part to the status of public thinking at which he was, whether consciously or not, directing his analysis— the Wisconsin public needed to be persuaded there were too many deer, so he would persuade them, using the history, theory, and threat of irruptions. Cuttings, fire, openings, and plantings took their place alongside buck laws, predator control, and refuges in Leopold's theory of irruptions as "predisposing causes." With a herd already in process of irruption, cuttings or plantings had

the same effect as artificial feeding: to postpone the ultimate inevitable starvation, in the meantime allowing the
deer to annihilate reproduction of the preferred browse
species. Leopold undoubtedly thought herd reduction in
threatened areas was so imperative that he could not risk
discussing possible techniques of habitat management,
even for use in the future, lest his opponents seize on
them as an argument against the necessity of reduction.
In point of fact, north country residents did argue for
more and better-regulated cutting and planting as *alternatives* to reduction.

The picture Leopold drew of Wisconsin's deer prospects
in his speeches and articles during the decade of the 1940s
was thus a dismal one. Wisconsin could reduce its herd
immediately and drastically by at least 60 per cent or wait
until the deer ruined their entire browse supply and with
it the future forest and then starved. The buck law and
strict enforcement, predator control, refuges and closed
areas, artificial feeding, forest cuttings, and fire protection—all of these measures by which Wisconsin citizens
had built up their herd to its generous proportions—
would be of no avail to preserve such herd numbers for
the future. At every opportunity he related how the Kaibab herd had destroyed its range and then itself and how
the Wisconsin herd was doing the same. It is not difficult
to see how some may have felt that Aldo Leopold was
conducting a vendetta against the deer.

With *Save Wisconsin's Deer* and other publications
and people ready to jump on him for the slightest con-
tradiction or evidence of uncertainty in his public pro-
nouncements, Leopold was practically "locked in" to his
irruption hypothesis and its implications for the future.
But even as he completed his "Survey of Over-Populated
Deer Ranges" he was having difficulty accounting for the
apparent failure of the Wisconsin herd either to continue
increasing at the rate projected by the Kaibab model or
to peak and then starve down dramatically, as he had
publicly predicted it would. He had yielded on the bounty
issue, as we have seen, in part because predators seemed
already to have responded to and trimmed the excess deer,
but he must have been uncomfortable with such an in-

terpretation. In early 1947, after his "Survey" had been submitted for publication, he sent a memo to the other commissioners explaining that he had now come to two new conclusions about the situation in Wisconsin.

The first was that all the unavoidable propaganda about "too many deer" had loosened the public conscience about hunting laws, with the result that deer hunters were now illegally killing almost as many antlerless deer as would have been taken under an any-deer law. His assertion was borne out that spring, when workers surveying deer damage to forest reproduction tallied all deer carcasses found and determined that there had been at least 45 deer killed illegally (and left lying in the woods) for every 100 deer taken legally in the central forest area. In maintaining a buck law despite a known overpopulation of deer, the state was thus unwittingly contributing to the ethical degradation of its hunters, Leopold pointed out. Thus was revealed the same natural law of predation as research was discovering in animals: "When there is a surplus, something is going to remove it; if one form of predation fails another takes over; if both fail, starvation steps in and finishes the job. This law operates through political as well as biological channels, and it involves both physical and ethical wastage."[44]

His second new conclusion was expressed in a uniquely Leopoldian phrase, "downhill equilibrium." Because of the heavy illegal kill of does, the herd was apparently "somewhere near stabilized"; that is, it was neither increasing as fast as would be expected under a buck law nor starving down as dramatically as would be expected if it had already peaked. Yet the range was still overstocked

44. "Memo for Commissioners on the Present Status of the Deer Herd," typescript, c. Jan. 1947, 3 pp., LP 2B10; and (for the quote) "Adventures of a Conservation Commissioner," p. 6. In one area where deer were heavily concentrated in Jackson County a special dead-deer survey after the 1947 season turned up 22 illegally killed deer per square mile. Systematic dead-deer surveys in eastern Jackson County from 1956 to 1964 indicated a combined illegal kill and crippling loss of from 2.4 to 6.5 deer per square mile, or from 35 per cent to 150 per cent of the registered legal harvest. See Bersing, *Century of Wisconsin Deer*, p. 251.

and hence carrying capacity was diminishing. This was a dangerous situation because the public, seeing little change in deer numbers or sex ratio and unable to perceive changes in vegetation, assumed everything was all right. "They do not realize," wrote Leopold, "that under this 'Downhill Equilibrium' our browse will ultimately be killed off more completely than it would under an out-and-out irruption of the Kaibab type." The same situation, he believed, existed in Pennsylvania and Michigan. There were certain similarities with his 1941 conjecture about "pure" ranges with a high percentage of palatables, such as the Kaibab, being more susceptible to violent irruptions than the "dilute" ranges of the Lake States, which offered an array of inferior buffer foods, the conjecture that had given rise to William Feeney's concern about "a prolonged struggle of mediocrity" in northern Wisconsin. Such hypotheses, if he had been able to work them out in research, might have led him to a deeper understanding of deer productivity as related to range quality, his quest ever since "Southwestern Game Fields."

In the absence of such research Leopold had difficulty, much as he had had in the 1920s, interpreting the effect of environmental change on deer numbers. After the split season of 1943, as we have seen, he had projected a state-wide carrying capacity of 200,000 deer if the herd were reduced promptly. In his explanations of illegal kill and downhill equilibrium he seemed to imply that the herd had already been reduced substantially, although not enough to bring it in line with carrying capacity. Yet the U.S. Fish and Wildlife Service in 1947 estimated that there were 791,000 whitetails in Wisconsin, and during the 1950s and 1960s the state would continue to carry well in excess of 500,000 deer.[45]

45. All figures on total population are, of course, estimates. After having been burned for the 1943 "guess" of 500,000, the conservation department refrained from making official estimates of total population. Even hunter kill figures were tentative until compulsory registration of deer at field checking stations was instituted in 1953. As kill registration, pellet count, and track count techniques were refined in the early 1960s, the department began making systematic population and density estimates by management areas, but even these were not pub-

There were two clusters of factors that could account for the state's capacity to carry so much larger a herd than Leopold projected, without peaking and starving down: 1) a rapid extension of deer range in central, southern, and eastern Wisconsin, together with greater productivity of the deer on the newly acquired range; and 2) the recuperative capacity of the northern range, new trends in forestry, and the adaptability of the northern herd to changing conditions. The first he did not clearly foresee; the second he underestimated. Yet, curiously, Leopold had observed, faithfully recorded, and then puzzled over elements of the emerging situation even as long ago as his game survey of Wisconsin in 1929. That he did not integrate them fully into a new interpretation of changing environmental conditions and deer herd response may be testimony to the hold which the theory of irruptions had on his thinking in the 1940s, as well as testimony to the lack of fundamental research on deer in Wisconsin and to the difficulty of projecting the direction and rate of ecologic change from scattered observations.

In his 1929 game survey of Wisconsin and in his *Report on a Game Survey of the North Central States* (1931), Leopold used a map on which he showed the major range boundary of whitetails in Wisconsin as including generally the northern counties, but he also showed blocks of deer range in the central sand and peat area and along the lower Wisconsin River and indicated a number of isolated remnants, outposts, and plants elsewhere in the state (Figure 4). In the published report he quoted Shiras on original scarcity of deer in the north and the following from Bogardus:

> It is often supposed that it (the deer) likes best to range in the vast forests, but I believe that to be a mistake. Deer are most fond of country in which there are belts of timberland and brush *interspersed with prairies and savannahs.* (Leopold's emphasis, p. 194)

lished and would not in any case admit of direct comparison with the 1943 estimates of similar areas. Over the state as a whole, the department estimated a total population of 600,000–650,000 deer for fall 1974 (James B. Hale, letter to author, 19 June 1974).

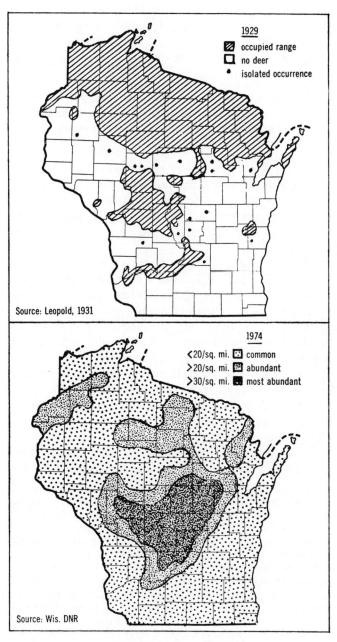

1929
occupied range
no deer
isolated occurrence

Source: Leopold, 1931

1974
<20/sq. mi. common
>20/sq. mi. abundant
>30/sq. mi. most abundant

Source: Wis. DNR

4. *Wisconsin Deer Range.*

Thus he concluded that the central part of the region was the qualitative center of the original deer range. Although the deer had shrunk northward with settlement, he observed that they were now partially recapturing their former southern range:

> Large southward encroachments are taking place in Wisconsin, and probably in the other Lake States. Some of these extend into the very edge of heavy farming districts.
> (p. 194)

Yet in his Wisconsin survey report, where he estimated total present and potential range for various species, he indicated a present range for whitetails of 10,100,000 acres, or 28 per cent of the state, and *no* potential increase. (The estimated range as of 1974 was over 20 million acres.)

His own "shack," of *Sand County Almanac* fame, was in the south-central region where deer were beginning to extend their range during his lifetime. The county was, in fact, one of the first two counties opened for bow-and-arrow deer hunting in 1934, at a time when none but the northernmost counties were opened to rifles, and Leopold acquired the shack in 1935 in part to serve as a base for archery hunts. By the early 1940s, the gun season had been extended to the central counties. One of the ironic consequences of Leopold's successful fight for herd reduction in 1943 was an excessive reduction near his own farm, 90 per cent he estimated.[46]

In his "Survey of Over-Populated Deer Ranges" Leopold puzzled over a phenomenon he termed *drift*. He

46. See AL to Asa K. Owen, 31 Aug. 1944, and AL to W. F. Grimmer, 29 June 1944, LP 2B10. In his capacity as a commissioner, Leopold wrote Grimmer of the conservation department to inquire about the factual basis for the 1944 congress vote for a buck season in the area: "I have already told you that in my own locality the reduction has been excessive, and I will have to be shown that the rest of these two counties is different before agreeing to any kind of a season." In the early 1970s, one of the largest herds in southern Wisconsin was wintering in the vicinity of the Leopold property and causing severe damage to vegetation, but citizen revolt against the variable quota system resulted in such heavy cut-backs on party permits that adequate reduction of the herd was impossible.

noted that "drifting deer began to show up on southern Wisconsin farms about 1930, which is also the time when the northern Wisconsin herd began to pyramid its numbers. . . . No similar drift is known to have occurred before 1930, when the northern herd was low." The presumptive evidence might seem to indicate that deer dispersed from a congested area as a result of population pressure. But Leopold's whole theory of deer irruptions was predicated on the "fact" that deer could tolerate extraordinary population densities and they would starve rather than move. As he put it, "Deer irruptions are possible because deer . . . differ from most other mammals, and from most birds, in social tolerance."

It is possible that this observed failure to disperse is more characteristic of mule deer than of whitetails, and of whitetails confined to northern yards by deep snows than of whitetails farther south or in milder weather. As for the supposed social tolerance of deer, subsequent research has revealed that density per se apparently *does* have adverse effects upon a population, such that average life span may be shortened and sexual maturity of females delayed. Autopsies of bucks found in some Wisconsin yards show death from internal injuries, presumably from fighting induced by psychological tensions in congested areas.[47] But such things were not understood in Leopold's day. "How to explain drift in so tolerant a species as deer," Leopold mused, "is still an unsolved riddle." Today, one suspects, he would focus on the newly habitable range into which the deer were moving and consider their occupation of it quite natural, deer psychology notwithstanding.

Whether or not deer dispersed and drifted southward as a result of density-induced stress is quite beside the point. According to Leopold's own 1929 survey, deer were already present in pockets in the south. They had only to multiply and spread from there in response to increasingly favorable environmental conditions. Many acres of central, southern, and eastern Wisconsin once cropped,

47. Burton Dahlberg, personal communication. See also A. Starker Leopold, "Too Many Deer," *Scientific American*, 193:22 (Nov. 1955), 101–4.

grazed, drained, or burned had been allowed to revert to natural woodland, brush, and marshland, like Leopold's own sand county farm. Milder winters, better soils, longer growing seasons, and abundance of agricultural foods all combined to allow greater herd productivity. By the 1960s research would reveal that the reproductive rate (fawns per doe) was twice as high in the southern herd as in the northern. Few game technicians realized even in 1961, when a harvest goal of 75,000–100,000 deer per year was projected into the future, that in half a decade the upper figure would be left far behind (there were 136,000 in 1967), largely as a result of further expansion in range and greater productivity in the southern part of the state. Waupaca County in east-central Wisconsin, for example, which Leopold did not indicate as having any deer at all on his 1929 map, in 1968 registered among the highest total legal kills in the state; it sustained a known mortality from hunting and highway losses alone of more than 20 deer per square mile of range, apparently without exceeding the natural increase for the year.[48]

It was thus in the southern half of the state that the real "irruption" was occurring in the 1940s. But Leopold, with his emphasis on the problem areas up north, on high populations, overbrowsing, and starvation winters—in short, on the symptoms of irruption—missed the significance of what was happening in his own backyard. Deer numbers, to be sure, were not yet noticeably high in the south nor was there much evidence of browse damage, except in a few areas of central Wisconsin such as Saddle Mound, Necedah, and the CWCA lands. But even the Kaibab herd, if he had thought about it, started with only a few thousand animals, and at the George Reserve in Michigan, another of his favorite sample irruptions, there

48. See Daniel Owen, "Deer Explosion in the South," *WCB*, 32:5 (Sept.–Oct. 1967), 14–15; and George F. Hartman, "Wildlife Management Notes," *WCB*, 33:6 (Nov.–Dec. 1968), 11. Compare Burton L. Dahlberg and Ralph C. Guettinger, *The White-Tailed Deer in Wisconsin*, WCD Technical Wildlife Bull. No. 14 (1956), p. 236: "Regardless of game and forest management favorable to deer which may be anticipated, the trend in deer numbers for the next two and possibly three decades will be down."

had been but four does and two bucks to start with. He had identified the southern half of the state as the qualitative center of the presettlement deer range; as early as his manuscript on the Southwest and again in his proposed deer study of 1941 he had regarded the effect of range quality on productivity, health, and carrying capacity as one of the most promising leads for research. Yet during the deer debates of the 1940s neither he nor anyone else talked of these things.

Just as Leopold underestimated the positive effect of habitat changes on deer populations in the south, so he overestimated their negative effect in the north, while he urged immediate herd reduction to prevent still greater degradation of the forest. He had warned publicly about the gradual replacement of palatable species by unpalatable ones as a result of overbrowsing, and to his students at the university he also explained the phenomenon of crown closure, which was a natural stage in the life cycle of a forest. "Deer habitat is good," he maintained, "in proportion to the diversity of its soils (and plant associations) and the diversity of its forest age–classes."[49] In Wisconsin, as in many other states, deer populations had irrupted because of a lack of diversity in forest age–classes; that is, because of a favorable stage in forest succession over virtually all the northern cutovers at the same time, as a result of newly effective fire protection in the early 1930s. The irruptive herd in turn, through pressure of overbrowsing, was impairing the diversity of plant associations. Highly palatable white cedar, hemlock, yew, white pine, red maple, alternate dogwood, yellow birch, and a host of herb and berry species were being run out by deer, and their place was being taken by increasing quantities of alder, aspen, hazel, white birch, soft maple, and balsam, all regarded as low palatables or starvation foods. The decline in carrying capacity, Leopold explained, would become even more severe as the young forests closed their crowns, thereby inhibiting reproduction of sun-loving understory browse. He knew that crown closure would occur whether or not there was overbrowsing by deer. But in his public statements during the 1940s he stressed over-

49. "White Tail Deer," p. 5.

browsing rather than crown closure because he was more concerned about what deer were doing to forests than about what forests were doing to deer. His prognosis of vegetational change in the forests of northern Wisconsin was essentially correct, but what he did not foresee was the extent to which both men and deer would adapt to the new regimen.

In particular, he did not comprehend the significance of the changing role of aspen in the Wisconsin forest economy. Curiously, it was the year Leopold left the Forest Products Laboratory (1928) that scientists there developed the neutral sulfite semichemical process that was to make possible the utilization of aspen and other second-growth hardwoods for pulp. By the 1930s more than a third of the state's forest acreage was dominated by the aspen–birch type, but it was not until the acute shortage of conifer pulpwood and the rising demand for paper products during World War II that aspen began to be accepted as a pulp species. By 1956 it had become the major commercial species in Wisconsin in terms of volume used, accounting for more than a third of the timber cut that year, and the balance of timber production in the state had shifted dramatically from sawlogs to pulpwood. Many foresters now sought to perpetuate aspen, birch, balsam, and other pioneer species in the forest succession, which could be grown to pulpwood size in as little as twenty-five years, and they simply forgot about far more valuable species of the mature forest such as white pine, white cedar, hemlock, and yellow birch, which were "uneconomical" because they would require a hundred years or more to grow to sawlog size, if indeed they were able to reproduce.[50] Even if he had foreseen the value of aspen as the fastest-growing of the pulpwood species, Leopold,

50. See Robert N. Stone and Harry W. Thorne, *Wisconsin's Forest Resources*, Lake States Forest Experiment Station, Station Paper No. 90 (Aug. 1961). There is no mention of deer damage to forest reproduction in this report, perhaps because it is concerned with the forest resource as it is, not as it might be. The best treatment of forest ecology and changes in vegetation is John T. Curtis, *The Vegetation of Wisconsin: An Ordination of Plant Communities* (Madison, 1959).

as the son of a desk maker and as an ecologist, would still have recoiled from it as he did from the spruce mania of the Germans, the monotypic jackpine plantations of the CCC, and the very idea of wholesale conversion to "cellulose forestry" and short rotation expedients.

Deer adapted as rapidly as foresters to the change in forest composition. By the 1960s deer were getting along all right on balsam and hazel, thought in the forties to be the end of the line in a starvation diet, and they were thriving on aspen, now considered the bread-and-butter browse species. The white cedar yards, which had seemed absolutely essential for winter survival, continued to decline. Hence deer did not yard as much as formerly; but, although some might die of exposure in severe weather, Wisconsin deer by the sixties proved their ability to winter successfully on current cuttings in the vast blocks of jackpine planted by the CCC. During the severe winters of the late sixties as much as half the food of deer in the north was supplied by cut browse, mostly from pulpwood logging operations. Deer normally eat more than a hundred different species of trees and shrubs in Wisconsin. When their preferred food species disappeared, they gradually turned to whatever was available.

Ecologists, including Leopold's son Starker, have suggested that deer probably evolved in an environment subject to frequent disturbance, such as by fire, and as a result are able to adapt to a wide range of transitory and unstable conditions.[51] Aldo Leopold sensed as much back in 1918 when he argued against the "wilderness fallacy" by pointing out how deer had been able to take advantage of many environmental changes wrought by the advance of civilization. In later years he came near suggesting on several occasions that deer, like aspen, were a weed species. Yet he did not anticipate the extent to which deer would adapt to the new forest economy.

Aldo Leopold was not alone in failing to appreciate the

51. See A. Starker Leopold, "Adaptability of Animals to Habitat Change," in F. Fraser Darling and John P. Milton, eds., *Future Environments of North America* (New York, 1966), 66–75.

southward spread of deer, their greater productivity in the central and southern portions of the state, and their adaptability to changing conditions in the northern forest. No one at the time understood what was happening, though a few like Leopold glimpsed pieces of the pattern and puzzled over them. The deer researchers on the Pittman–Robertson project were preoccupied with surveys in the 1940s, under pressure of demonstrating urgent need for herd reduction. It was not until the late forties and early fifties that they moved toward the more fundamental research on life histories and habitat relations that is reported in their *White-Tailed Deer in Wisconsin* (1956), and that of course was but a beginning.

So once again, as in his analysis of the deer–environment equation in the Southwest two decades earlier, Aldo Leopold had many of the ingredients for an interpretation that could have accounted for the fortunes of deer and forests in Wisconsin, but he had difficulty integrating them in a comprehensive ecological analysis. As in the Southwest, he may have been inhibited by his inability to get into the field for sustained observation and research, owing not only to wartime gas rationing and the postwar crush of returning students but also to other demands on his time, not the least of which was his service as conservation commissioner, and to his increasingly frequent bouts with ill health. Yet he did make a number of inspection trips to problem areas in the north, and he had regular contact with researchers who spent most of their time in the field. It is difficult to escape the conclusion that another important reason for his difficulty with ecological interpretation was his preoccupation with the phenomenon of irruptions as related to the public problem and the necessity for immediate herd reduction.

1948: Denouement

Aldo Leopold died on April 21, 1948. It was the time of year when the conservation department, the conservation congress, and the commission began preparing for their annual round of deer debates. There was reason to believe that 1948 might finally be a year of herd reduction, for the conservation congress seemed to be shifting gears preparatory to a change of direction. Leopold might have died with this hope. Yet he would have been under no illusions that one year of herd reduction or even a series of them would in any real sense "solve" the problem. It was too complex, too dynamic for solution. The best that could be hoped was that researchers, administrators, policymakers, and the public would begin converging and continue to move in a positive direction. By the time of his death, Aldo Leopold might have had the satisfaction of believing that Wisconsin was beginning to move constructively to deal with the deer problem and that he more than anyone had helped to overcome inertia and to point the way. Yet in the immediate aftermath of his death confusion reigned for yet another year.

In the winter of 1947–1948 the leaders of the conservation congress, who had been cooperating with the department deer researchers since the earliest Pittman–Robertson surveys, decided to appoint a committee of the congress to conduct its own field examination, in hopes that the congress might give more credence to a report from its own members. A number of newsmen wrote scornfully of yet another study group fancying it knew better than the experts. "Why not let game management do its job?" queried Russell Lynch, sports editor of the *Milwaukee Journal*. Aldo Leopold, on the other hand, had written to Clarence Searles, chairman of the congress, "No doubt we should have tried this committee business three or four years ago." He undoubtedly had recalled the impression made on northern members of his own citizens' deer committee when he took them to the field in the winter of 1943. By some quirk of fate, the three winters in which newly appointed deer survey committees were in the field (1943, 1945, and 1948) happened also to be the three most severe starvation winters in northern Wisconsin during that decade. The initiates, on each occasion, were shocked into awareness. Early in April the deer committee submitted to the executive council of the congress its unanimous recommendation for immediate reduction of the herd.[52]

Aldo Leopold may never have seen the committee's report, for it was not officially issued until April 29, eight days after he died. The 1948 congress, convinced by a committee of its own members that herd reduction was indeed imperative, voted 37 to 33 for an any-deer season. With the congress and the department in agreement, the commission finally voted 4 to 2 to adopt the any-deer recommendation.

The summer of 1948 saw incumbent Governor Oscar Rennebohm in a hot race for reelection. Although he had never before challenged a conservation commission order, the governor vetoed the deer decision, citing the "volume of protest from the areas most directly concerned" and Joint Resolution No. 54,S, of the 1947 legislature. "The Conservation Congress," he wrote in his veto message to the commission, "represents sportsmen and conservationists from all 71 counties. The Legislature, of course, represents all the people of the state." Political observers did not doubt that the governor made political hay by his veto. "In northern Wisconsin," one reporter noted, "voters are fully capable of making a selection for governor according to his stand on deer."[53]

At the August commission meeting, W. J. P. Aberg, who had voted in July with the majority for an any-deer season, gave up in disgust and changed his vote to buck, saying:

> We know that a buck season is not the right thing, but if
> it is what the people want let's give it to them, but let them

52. R. G. L., "Right Off the Reel," *Milwaukee Journal*, 22 Feb. 1948; AL to Clarence A. Searles, 9 Feb. 1948, LP 2B10. See also "Unanimous Report of the Wisconsin Conservation Congress Deer Committee," typescript, submitted 29 April 1948, 20 pp., copy at U.W. Dept. of Wildlife Ecology.

53. Oscar Rennebohm to State Conservation Commission, 24 July 1948, in commission minutes, 10–11 Aug. 1948; and John Wyngaard, "Government and Politics," *Green Bay Press-Gazette*, 10 Sept. 1948. The governor's letter was somewhat contradictory, on the one hand calling for cognizance of Jt. Res. 54,S, according to which the commission was to have sought approval from the county board of each county affected by an

know we're temporizing. We're at the end of a rope as far as deer management is concerned. Let's let the good Lord take care of it in the future.

Deadlocked 3 to 3, the commissioners postponed their decision yet another month while department administrators continued their efforts to prepare a plan for special refuge areas that might meet with the governor's approval. Russell Lynch, trying his utmost to get Aberg to reconsider his stand, invoked the spirit of Aldo Leopold in his *Milwaukee Journal* column the day before the September commission meeting, under the head, "A 'Seven Man' Commission Will Debate Deer":

> Surely, if there be any return after death, the spirit of Aldo Leopold will be there, sitting beside his friend, Bill Aberg, with whom he fought, shoulder to shoulder, for sound game management through the years. . . .
> If Leopold had been alive, to accept any small gain with his infinite understanding and patience, the commission would have submitted a substitute plan at its Milwaukee meeting last month. Without him, Aberg gave up in disgust and the commission deadlocked.
> Perhaps, if the spirit of Aldo Leopold sits at the table Friday in Eagle River, the story will be different.

Aberg did not reconsider. To break the stalemate, two more of the remaining any-deer advocates on the commission changed their votes reluctantly to buck. "Convictions," one of them conceded, "are not worth a plugged nickel if by forcing them on a reluctant, bitter minority, harm can come to the commission, the department, and conservation program":

> And, inevitably, the program suffers set-backs, and I do have a fear of the political backwash—so, with a knowledge of the past, I very reluctantly, and even apologetically, in deference to those conservation leaders who have devoted so

---

any-deer season, and on the other hand asking that the commission simply satisfy the protesters by giving representatives of the northern counties an opportunity to indicate areas which should be designated as refuges.

much personal effort to gain deer-herd reduction, and have had strong support from many splendid people all over the state, under the circumstances—and circumstances I do not favor—I vote 'Yes' for a forked-horn buck regulation.[54]

* * *

Talking about the "Adventures of a Conservation Commissioner" at a wildlife conference little more than a year before his death, Aldo Leopold shared some of the conclusions he had reached after a lifetime of setting up commissions and half a decade of serving on one. A good commission, he observed, could prevent a conservation program from "falling below the general level of popular ethics and intelligence," but no commission could raise its program much above that level, except in matters to which the public was indifferent. "Where the public has feelings, traditions, or prejudices," he said, "a Commission must drag its public along behind it like a balky mule, but with this difference: the public, unlike the mule, kicks both fore and aft." In obvious reference to the deer problem in Wisconsin he continued: "An issue may be so clear in outline, so inevitable in logic, so imperative in need, and so universal in importance as to command immediate support from any reasonable person. Yet that collective person, the public, may take a decade to see the argument, and another to acquiesce in an effective program."[55]

54. Aberg is quoted in R. G. Lynch's column, "Maybe I'm Wrong," *Milwaukee Journal*, 13 Aug. 1948. "Aberg either was disgusted at the governor's refusal to support sound deer herd management," Lynch suggested, "or he was feigning disgust as an excuse for complete surrender to a belligerent minority." Leopold's spirit is invoked in Lynch's column of 9 Sept. 1948; the commissioner's statement, by Arthur Molstad of Milwaukee, is in the commission minutes, 9–10 Sept. 1948.
55. "Adventures of a Conservation Commissioner," p. 1.

Epilogue

What Happened in Wisconsin?

Finally in 1949 Wisconsin had an antlerless season, and in 1950 and 1951 seasons on any deer. The harvests averaged over 150,000 deer per year, as compared with 128,000 in 1943, and in each season Wisconsin led the nation in the number of white-tailed deer killed. Ernest Swift warned that a third liberal season would be one too many, and severe public reaction after the 1951 kill bore him out. There followed a retrenchment to five years of buck seasons, the first three with kills lower than in any year since the 1930s. From 1957 to 1960 the kill was boosted by the issuance of permits for any group of four hunters to take one deer of either sex for the party in addition to a buck for each individual, but poorly distributed harvests resulted in prohibition of the party-permit system by the 1961 legislature. The same legislature, however, finally gave the commission its long-sought authority to conduct controlled hunts, beginning in 1963. The new system, known as the variable quota plan, divided the state into management units and provided for a general buck season in all open units and, in addition, a specified number of party permits for a deer of either sex in each unit where deer were in need of reduction. Except that it applied to does as well as to bucks, the variable quota feature was essentially what Leopold had proposed as early as 1918.

Wisconsin by the mid-1960s was considered to have one of the best systems of deer herd management in the nation. The state was divided into seventy-seven management units, roughly according to habitat characteristics, and they served as the basis for zoned harvest under the variable quota system. Standardized censusing procedures gave an objective measure of relative deer density from year to year in the various units, and compulsory field registration of kills provided credible statistics concerning the harvest. By substituting facts for "guesses" and demonstrating through a succession of harvests of over

100,000 deer that the state was able consistently to maintain both herd and harvest at a level higher than anyone
would have dreamed in Leopold's day, state conservation
officials won at least that modicum of public acquiescence
they needed to continue scientific herd management.

Even in the halcyon years of the 1960s, however, one
could hardly find a wildlife manager in Wisconsin who
did not believe personally that there were still too many
deer in many parts of the state; and the general public
thought there were still too few. The variable quota system took the feast and famine out of deer hunting and
smoothed out the harvest curve for the state as a whole;
but for any given unit, especially in the north, the harvest was just as the term implied, variable. Under certain
Wisconsin conditions, a forest area that was clearcut
could change from poor deer range to good and back
again in as short a time as ten years. The effect of a severe
winter is even more immediate, and not only increases
starvation losses but also reduces fawn production. Deer
hunters who are in the habit of returning to a familiar
camp year after year and who do not understand forest
succession and the environmental requirements of deer
are apt to blame past high quotas when deer numbers
decline in their area, instead of using present high quotas
in other units as a key to new hunting grounds where
deer are abundant. Thus, as a result of environmental
changes, hunting habits, and the vagaries of weather,
many areas remain underharvested and conservation officials worry about too many deer, while hunters complain of too few deer and the evils of the variable quota.

Occasionally latent discontents erupt and interfere with
management, as in 1969–1970. The winter of 1968–1969
was a severe one in northern Wisconsin, with more than
125 days of snow cover. Deer populations and productivity declined and the quotas were reduced accordingly,
but not near enough to satisfy many residents of the
north. During the 1969 season 97,987 deer were killed,
and irate sportsmen in the north rose in revolt against
the variable quota. County board supervisors, ever sensitive to public opinion, readied resolutions calling for a
ban on the party permit in order to protect the does and

fawns. When the state assembly returned for two weeks in January 1970 to deal with a backlog of old business, one of the first measures they considered was a new bill introduced by an assemblyman from Vilas County providing for a one-year moratorium on the variable quota system. It passed, 86 to 13.

Although no action was taken by the state senate and the moratorium did not become law, the department had been served warning that it would have to tread lightly for a time until opposition subsided. Over the years game managers in the department had acquired a certain philosophical detachment from which they could view such northern discontent as perennial. Thus, when the 1970 conservation congress insisted that the number of party permits recommended by the department be cut by 50 per cent, the department, gambling on an easy winter, gave in without even taking the issue to the Natural Resources Board (successor to the conservation commission). The 1970 harvest was only about 70,000 deer, but that winter the state lost an estimated 50,000–60,000 deer to starvation—probably more than in any winter since 1943, when Aldo Leopold took his citizens' deer committee to the field on snowshoes to show them the consequences of too many deer.

The somewhat lower harvests of 1970 and succeeding years were caused in large part by a series of severe winters, but the impact of severe winters was exacerbated in the north by long-term environmental changes, which lowered carrying capacity and left high deer populations subsisting on an inadequate supply of marginal-quality forage. The astonishing increase in deer populations and productivity in the central and southern portions of the state could not entirely mask what was happening in the north. Adaptable as deer were to changing conditions, by the 1960s there were many areas in the north in which there was simply no food within reach. Although attention in the forties had been focused almost exclusively on the winter yards, the increase in pulpwood logging now supplied adequate cut browse during most winters, and game managers were beginning to consider the availability of food on the summer range as the limiting factor for deer. Crown closure had occurred over much of the north as Leopold had warned it would; it was particularly serious in stands of low-quality hardwoods from which aspen had been cut selectively, because the remaining trees not only shaded out reproduction of aspen and other browse species but were too valueless to consider cutting. Some areas of previously high deer populations, such as Iron County, became virtual wildlife deserts. As early as his report on the Rockford deer area in 1937, Leopold had alluded to the problem of summer range; but whereas he had been concerned principally about maintaining vegetational diversity, game managers were now experimenting with ways to get aspen and more aspen.

As the interest of both deer managers and foresters converged on aspen and other species useful for both pulpwood and browse, the state conservation commission finally began to encourage the formulation of policies to integrate management of deer and forest. It was agreed that since most of the deer range was commercial forest land, the most important management for deer consisted of intensive forest management, involving continuous cutting and subsequent regeneration. Beyond encouraging maximum timber sales and development of new markets for wood products, there were also possibilities for redirecting forestry operations to meet requirements for deer habitat. They included proper timing and distribution of cuttings; use of various cutting methods, such as clearcutting of aspen, clearcutting jackpine and lowland conifers in strips, shelterwood cutting of white and Norway pine, and selection cutting in northern hardwoods; discing, bulldozing, burning, and other site preparation techniques; and maintenance of the proper interspersion of forest openings.[1]

1. See Wisconsin Conservation Department, "Deer-Forest Interrelationships in Forest Land Management," mimeographed, 1962, 60 pp. The conclusions of this report were adopted as commission policy 21 Sept. 1962. See also Laurits W. Krefting et al., "Research for Deer Management in the Great Lakes Region," a contribution of the Great Lakes Deer Group, mimeographed (Dec. 1964), 73 pp.; Keith R. McCaffery and William A. Creed, *Significance of Forest Openings to Deer in Northern*

Such practices, although incorporated into official policy, proved difficult to inaugurate even on state-owned lands. Many foresters, on public as well as private lands, do not look kindly on the prospect of creating or maintaining openings for wildlife. Nor are they particularly happy to encourage a proliferation of brush beyond what is needed for forest reproduction. They would rather not have to worry about deer damage and deer hunters and they know how to avoid both—by planting vast unbroken tracts to monotypes and using herbicides to kill undesired brush. Sportsmen complain bitterly that modification of cutting practices and other habitat improvements in the interest of deer are not proceeding nearly fast enough, but on the other hand they insist that the increase in clearcut pulpwood sales means the state can support a larger herd.

That discussion, policy, and at least the beginnings of action by the 1960s revolved around questions of habitat management indicates a certain fulfillment of the game management sequence of ideas identified by Aldo Leopold some three decades earlier. Wisconsin can probably continue indefinitely to maintain a level of cellulose and venison production equal to that in the 1960s and, with more intensive management, higher. Leopold might well be given a large share of the credit for this development, but he would not be satisfied with the situation.

There is as yet limited understanding of ecological interrelationships among the general public and little appreciation, at any level, of environmental considerations going beyond short-run forest economics and deer hunting. The issue is still too few deer or too many, rather than adjustment of the herd in line with the requirements of land health. A recent survey of Wisconsin hunters revealed that almost half still thought doe deer should never be hunted and over 70 per cent favored the use of bounties to control predators such as foxes. Among nonhunters there seems to be mounting sentiment against the whole

---

Wisconsin, DNR Tech. Bull. No. 44 (1969); Frank Haberland, "We're Out to Grow Deer Food," *WCB*, 38:5 (Sept.–Oct. 1972); and James O. Evrard and Gerald A. Spoerl, "Versatile Aspen," *WCB*, 38:5 (Sept.–Oct. 1973), 11–14.

idea of hunting, with no thought for the ecological consequences of leaving deer herds uncontrolled. Despite more frequent calls for preservation of scientific and natural areas and even tag ends of "wilderness," there is little understanding of the impact deer may have on vegetational diversity in such areas, not to mention their impact on the rest of the forest. Since Leopold's day ecologists have learned more about other factors besides deer, especially the exclusion of fire, which may inhibit reproduction of many tree species once characteristic of northern Wisconsin, but people who do not understand the impact of overbrowsing by deer often are not attuned to other factors. Many public officials, commercial forestry interests, sportsmen, and even some self-styled environmentalists have been willing to acquiesce in perpetual pulpwood on much of Wisconsin's forest acreage, seemingly without even considering whether there is anything more a forest might be.[2]

As a symbol of what was and of what could be there is the former Menominee Indian Reservation, largest block of old-growth timber in the Lake States, which was used as a base datum of natural forest reproduction in the deer damage survey of 1947–1948. This area of white and red pine, hemlock, yew, sugar maple, basswood, yellow birch, and all the associated flora and fauna, has since the early years of the century been harvested on a sustained yield basis by conservative selective cutting. The annual growth increment in the 1960s was 218 board feet per acre, as compared with only about 100 board feet for most of the forest land in northern Wisconsin, an indication that economy as well as ecology and esthetics was sacrificed when the northern forests were leveled. The success of new natural forest reproduction has been directly attributed to the low density of deer, estimated at less than four per square mile in 1960, kept that low as a result of year-long unrestricted hunting by the Indians. Whites hunting with conventional methods under the usual game regula-

2. Results of the hunter survey are reported in Lowell L. Klessig and James B. Hale, *A Profile of Wisconsin Hunters,* DNR Tech. Bull. No. 60 (1972).

tions would not be able to keep a deer herd trimmed to such a low density.[3]

Yet, after termination of reservation status by the federal government became effective in 1961 and the area became a county, the Wisconsin Conservation Commission decided to place on Menominee County the same seasons and bag limits as on all the other counties of the state, holding that the Indians lost their tribal hunting rights upon termination. The effect of restricting the Indians' freedom to hunt, as a few employees of the conservation department predicted, was a dramatic increase in the deer population within a few years and serious damage to new forest reproduction. The age-long integrity, stability, and beauty of the Menominee forests may have been rescued by a Supreme Court ruling that the Menominee had in fact not lost their ancestral hunting rights and by an Act of Congress in 1973 that restored Menominee tribal and quasi-reservation status. One is left to imagine Aldo Leopold's reaction, had he been a member of the commission at the time of the decision to restrict the hunting of Menominee deer.[4]

Ecology and Ethics

All his life Aldo Leopold thought of himself as one of a minority. "For us of the minority," he wrote in the foreword to *Sand County Almanac*, "the opportunity to see geese is more important than television, and the chance to find a pasque-flower is a right as inalienable as free speech." And again, "We of the minority see a law of

3. See Clarence J. Milfred et al., *Soil Resources and Forest Ecology of Menominee County*, University of Wisconsin Geological and Natural History Survey, Soil Survey Division, Bulletin 85, Soil Series No. 60 (1967).

4. Max Morehouse and Robert J. Becker stated the case for more liberal hunting regulations in "Menominee County Deer," *WCB*, 31:4 (July–Aug. 1966), 20–21. The range of issues in the Menominee controversy extends far beyond the deer question, as revealed in Deborah Shames, ed., *Freedom With Reservation: The Menominee Struggle to Save Their Land and People* (Madison, 1972).

diminishing returns in progress; our opponents do not."
From this sense of speaking for and to a minority stems
much of the clarity and strength of *A Sand County Almanac*. Leopold's writing in the *Almanac* has a timeless
quality, philosophically detached from the problems of
the present. It offers a perspective from history and ecology, a way of thinking, to those who would accept it,
either now or in the future.

By contrast, Leopold's stance in the deer debates of the
1940s was rather uncharacteristic for him and probably
uncomfortable, for he was addressing the general public,
a majority of whom were not of his persuasion, and appealing for action in the immediate future. Deer management was the one big issue on which Leopold fought in
the public arena. It was a crisis, but there were other crises
during which he was content to remain behind the
scenes and work through key individuals, such as land
managers, administrators, policymakers, students. He was
a member of a public policymaking body, the conservation
commission, but his appointment was in large part a consequence of his earlier efforts to win herd reduction. To
understand why Leopold entered the public arena on
the deer issue one must understand his past experiences
with deer and the evolution of his own ecological attitude.
He was personally involved in the deer issue in a way that
virtually precluded philosophical detachment.

Once before, at the beginning of his career, he had tak-
en his message directly to the public. The marked contrast
between his fabulous success at organizing for game protection in the Southwest and his frustrating attempts to
win public support for deer herd reduction in Wisconsin
points up a vast difference in the character of the messages
he was trying to get across in the beginning and at the
end. The early message was a simple one of resource
supply: Deer and other game animals were in short supply relative to hunting demand, and Leopold was offering an attractive proposal to restore abundance. The
later message had to do with land health; it was premised on the welfare of the community rather than solely
of the species, and it was not framed in the market
terminology to which people were accustomed. To their

minds, Leopold was calling for a reduction in supply when what they wanted was more deer. Herd reduction was a means to the end of land health, but people who did not understand the end mistrusted the means.

Leopold occasionally referred to deer as one of the few "easy species" to manage, by which he meant that fluctuations in deer populations could be readily understood in terms of visible causes. Yet, deer ecology, some deer researchers would insist, is farther from being understood today than it seemed in Leopold's day. And deer management, as a public problem, involved vastly more than an understanding of deer ecology. Leopold himself did not appreciate all the ramifications in ecology and policy. But he saw very clearly, more clearly perhaps than any of his contemporaries, the ecological context in which the issues had to be formulated and the implications for public policy. What is more important, he was evolving for himself, and endeavoring to impart to others, a way of thinking ecologically about dynamic situations.

What was required, in Leopold's thinking, if normally functioning biotic communities like Menominee County, the Chihuahua sierra, and Huron Mountain were to be preserved and the disturbed lands of most of Wisconsin and the rest of the nation were to be restored to some measure of ecological integrity, was a transmutation of values on the part of the public as a whole. It had to be accompanied by "an internal change in our intellectual emphasis, loyalties, affections, and convictions" and reflect a consciousness of individual responsibility for the health of the land.

Leopold presented the land ethic as a product of social evolution, which in turn is a product of evolution in the minds of countless individuals. Through his own intellectual evolution, he paralleled and indeed advanced the development of ecological science. But it was a journey which each individual would nevertheless have to make for himself. Perhaps with his frustrations at the public response to the deer problem in mind, Leopold insisted that the evolution of a land ethic was an intellectual as well as emotional process. "Conservation is paved with good intentions which prove to be futile, or even dangerous," he warned, "because they are devoid of critical understanding either of the land, or of economic land use. I think it is a truism that as the ethical frontier advances from the individual to the community, its intellectual content increases."[5] Ecological understanding and ethical attitudes, moreover, would have to continue to evolve as the land organism itself evolved.

It is the land ethic—his conception of land health, or the philosophy of a natural self-regulating system, coupled with his assertion of individual obligation—that represents Aldo Leopold's most enduring contribution. Whereas to some the notion of a self-regulating system may be an invitation to fatalism, to Leopold the concept of a land ethic implied an obligation to help restore conditions that, according to the best available knowledge, may be conducive to land health. Hence he was concerned all his life not only with furthering ecological understanding but also with developing the tools and techniques of land management and criteria for applying them. Early in his career he had visualized the management arts and sciences developing to such an extent that it would be possible eventually to shape and maintain a controlled environment, but his experiences with deer convinced him that the land organism was too complex and dynamic ever to be fully comprehended or controlled and that management, however essential, was itself subject to the same hazardous consequences as the short-sighted actions it was intended to correct. His career thus seems an extended ironic dilemma. He was a man engaged in attempting to make adjustments in order to restore a self-adjusting system so that adjustments would be unnecessary. As he once expressed it, "In the long run we shall learn that there is no such thing as forestry, no such thing as game management. The only reality is an intelligent respect for, and adjustment to, the inherent tendency of land to produce life."[6]

5. "The Land Ethic," *A Sand County Almanac* (New York, 1949), p. 225.
6. [Review of Ward Shepard, *Notes on German Game Management, Chiefly in Bavaria and Baden*], *Journal of Forestry*, 32:7 (Oct. 1934), 775.

Epilogue

* * *

One of the last things Leopold wrote was a biographical tribute to a colleague in the Forest Service, C. K. Cooperrider, which was published in the same issue of the *Journal of Wildlife Management* that announced his own death. Cooperrider had devoted his life to researching problems of soil erosion and land use in the Southwest. As much as any other individual of his generation, Leopold felt, he had created the present ecological concept of man and land—the basis of a land ethic. "A new idea is, of course, never created by one individual alone," Leopold wrote. "A prophet is one who recognizes the birth of an idea in the collective mind, and who defines and clarifies, with his life, its meanings and its implications." [7] In that sentence, written in tribute to a friend, Aldo Leopold captured the essence of his own intellectual odyssey.

7. "Charles Knesal Cooperrider, 1889–1944," *Journal of Wildlife Management*, 12:3 (July 1948), 337–39.

Bibliographical Note

The major sources for this study have been the writings of Aldo Leopold, both published and unpublished; correspondence, reports, and other manuscript material in the files of Leopold, a number of his colleagues, and public agencies with which he was associated; discussions with members of the Leopold family, colleagues and students of Leopold, and current personnel of various agencies; news stories, scientific and technical articles and bulletins, and secondary historical works. This note provides an indication of general sources consulted during the study and some guides to further reading. It is not an attempt to relist all the items cited in the footnotes or to acknowledge the full range of uncited materials consulted by the author.

For published biographical essays on Leopold see Paul L. Errington, "In Appreciation of Aldo Leopold," *Journal of Wildlife Management*, 12:4 (Oct. 1948), 341–50; Ernest Swift, "Aldo Leopold," *Wisconsin Tales and Trails*, 2:3 (Fall 1961), 2–5; Roderick Nash, *Wilderness and the American Mind* (New Haven, 1967), pp. 182–99; and Susan Flader (with Charles Steinhacker, photographer), *The Sand Country of Aldo Leopold* (New York, 1973). The first chapter of the present study was originally published in slightly different form as "Thinking Like a Mountain: A Biographical Study of Aldo Leopold," *Forest History*, 17:1 (April 1973), 14–28. A full biography of Leopold with a bibliography of his publications is in progress by the author.

Leopold's major works include *Report on a Game Survey of the North Central States* (Madison, 1931); *Game Management* (New York, 1933); *A Sand County Almanac and Sketches Here and There* (New York, 1949); and *Round River: From the Journals of Aldo Leopold*, ed. by Luna B. Leopold (New York, 1953).

The most important source of unpublished writings, correspondence, and other manuscript material used in this study is the Aldo Leopold Papers in the University of Wisconsin Archives. The papers, deposited by the Leopold family, the

Bibliographical Note

University of Wisconsin Department of Wildlife Ecology, and many associates of Aldo Leopold, are included in Record Group 9/25/10, College of Agriculture, Wildlife Ecology (cited LP with series and box number). Although they span the years of Leopold's career, they are most complete after 1928 when he left the U.S. Forest Service.

Most of his Forest Service correspondence, memoranda, and reports remained in official government files where they are unindexed and extremely difficult to find. Records of the Washington office of the Forest Service (Record Group 95) have been consulted in the National Archives, Washington, D.C. (cited NA–RG95). Records of the Southwestern District, including documents pertaining to the Gila and other forest areas, are mostly at the Federal Records Center in Denver (cited FRC with container number), although a few files were consulted at the regional headquarters in Albuquerque and at various forest offices. Personnel files are at the Federal Records Center in St. Louis and records of the Forest Products Laboratory are at the laboratory in Madison. Some of the papers of J. Stokley Ligon are at the Conservation Library Center of the Denver Public Library.

For the Wisconsin phase of the study the voluminous and as yet unarranged records of the Wisconsin Conservation Department (WCD) were explored at the Archives of the State Historical Society of Wisconsin in Madison (SHSW). Minutes of commission meetings, reports of the Pittman–Robertson project, and a few miscellaneous records were consulted at the Department of Natural Resources (formerly WCD) in Madison. The files of W. J. P. Aberg were examined at his office in Madison, but they are now in the SHSW Manuscript Division. Also at SHSW are tape-recorded interviews with Aberg, Ernest Swift, and other conservation leaders and associates of Aldo Leopold. For legislative records there is the Legislative Reference Library at the state capitol, which also maintains extensive subject files of newspaper clippings. Other clippings, miscellaneous correspondence, bound volumes of key periodicals, and conservation department reports were generously loaned to the author by Walter E. Scott from his extraordinary private library of Wisconsin conservation history. Background reading in government reports and scientific and technical journals and bulletins was facilitated by permission to use the private libraries of Aldo Leopold and other professors at the University of Wisconsin Department of Wildlife Ecology.

Among the more helpful historical studies are Samuel P. Hays, *Conservation and the Gospel of Efficiency: The Progressive Conservation Movement, 1890–1920* (Cambridge, Mass., 1959); Samuel Trask Dana, *Forest and Range Policy: Its Development in the United States* (New York, 1956); James B. Trefethen, *Crusade for Wildlife: Highlights in Conservation Progress* (Harrisburg, Pa., 1961); and Roderick Nash, *Wilderness and the American Mind* (New Haven, 1967). Public policy issues, attitudes, and decisionmaking processes in nineteenth-century Wisconsin are brilliantly analyzed in James Willard Hurst, *Law and Economic Growth: The Legal History of the Lumber Industry in Wisconsin, 1836–1915* (Cambridge, Mass., 1964). The history of ecological science is told in W. C. Allee et al., *Principles of Animal Ecology* (Philadelphia, 1949); an early classic is Charles Elton, *Animal Ecology* (London, 1927). F. Fraser Darling writes of many of the individuals and areas mentioned in this study with an ecological attitude remarkably akin to Leopold's in *Pelican in the Wilderness: A Naturalist's Odyssey in North America* (New York, c. 1956). Donald Fleming cogently analyzes the writings of major intellectual leaders of the contemporary environmental movement, including Leopold, in "Roots of the New Conservation Movement," *Perspectives in American History*, 6 (1972), 7–91.

The nature and causes of environmental change in the Southwest are analyzed in Luna B. Leopold, "The Erosion Problem of Southwestern United States" (Ph.D. Dissertation, Harvard, 1950); James Rodney Hastings and Raymond M. Turner, *The Changing Mile: An Ecological Study of Vegetation Change With Time in the Lower Mile of an Arid and Semiarid Region* (Tucson, 1965); and Charles F. Cooper, "Changes in Vegetation, Structure, and Growth of Southwestern Pine Forests Since White Settlement," *Ecological Monographs*, 30:2 (April 1960), 129–64. Research on the Black Canyon deer herd is summarized by Robert H. Stewart in "Historical Background of the Black Range," 30 July 1962 (W–75–R–9, Work Plan 20, Job 7 Completion Report, New Mexico Department of Game and Fish), while the history of the Kaibab herd is told in John P. Russo, *The Kaibab North Deer Herd—Its History, Problems, and Man-*

agement, Arizona Game and Fish Department, Wildlife Bulletin No. 7 (July 1964). For a discussion of deer and wolves in the Sierra Madre see A. Starker Leopold, *Wildlife of Mexico: The Game Birds and Mammals* (Berkeley, 1959).

The best sources for the ecology of Wisconsin forests are John T. Curtis, *The Vegetation of Wisconsin: An Ordination of Plant Communities* (Madison, 1959) and Clarence J. Milfred et al., *Soil Resources and Forest Ecology of Menominee County*, University of Wisconsin Geological and Natural History Survey, Soil Survey Division, Bulletin 85, Soil Series No. 60 (1967).

The most useful works on the deer problem in Wisconsin are: Ernest Swift, *A History of Wisconsin Deer*, WCD Pub. 323 (March 1946); Burton L. Dahlberg and Ralph C. Guettinger, *The White-Tailed Deer in Wisconsin*, WCD Technical Wildlife Bull. No. 14 (1956); Otis S. Bersing, *A Century of Wisconsin Deer*, WCD Pub. 353–66 (1966); and the *Wisconsin Conservation Bulletin* (cited *WCB*), 1936-present. For a case study from another state see Paul Tillett, *Doe Day: The Antlerless Deer Controversy in New Jersey* (New Brunswick, N.J., 1963), an analysis by a political scientist with special attention to interest groups and other political forces.

For a classic in the literature of game management, with examples drawn largely from the author's extensive field experiences in Wisconsin, see Wallace Byron Grange, *The Way to Game Abundance: An Explanation of Game Cycles* (New York, 1949). Raymond F. Dasmann presents a textbook approach to principles of game management in *Wildlife Biology* (New York, 1964). Scientific research on deer up to the mid-1950s is summarized in Walter P. Taylor, ed., *The Deer of North America* (Harrisburg, Pa., 1956), for which Leopold was originally to have contributed a chapter. William Dasmann, *If Deer Are to Survive* (Harrisburg, Pa., 1971) is a recent more popular account of research on food and habitat requirements. For a sophisticated systems approach to wildlife ecology that incorporates many examples from recent scientific research on deer, see Aaron N. Moen, *Wildlife Ecology: An Analytical Approach* (San Francisco, 1973). On wolves see L. David Mech, *The Wolf: The Ecology and Behavior of an Endangered Species* (New York, 1970) and the monographs cited in Footnote 6 of Ch. 6.

One of the best post-Leopold statements of an ecological philosophy is in Charles Elton, *The Ecology of Invasions by*

Animals and Plants (London, 1958). See also Paul Shepard, *Man in the Landscape: A Historic View of the Esthetics of Nature* (New York, 1967); Rene Dubos, *The Mirage of Health: Utopias, Progress, and Biological Change* (New York, 1959); and Garrett Hardin, *Exploring New Ethics for Survival: The Voyage of the Spaceship Beagle* (New York, 1972). For a deft analysis of the process by which ideas or "images" grow and change see Kenneth E. Boulding, *The Image: Knowledge in Life and Society* (Ann Arbor, 1956).

Index

277

Costley, Richard, 117–78, 180
Cottam, Grant, 7n
Cowles, Henry C., 6
Coyotes: and deer, 93–96, 106, 111, 212–14; and wolves, 154, 212–15
Cramer, H. T. J., quoted, 238
"Crime of '43," 198–202, 204–5
Crisis, responsibility in, 162, 169, 172, 206–8, 229–30, 230n, 268
Crown closure, 253–54, 264
Culbertson, Victor, 77
Curtis, John T., 7n

Darling, Jay N., 27, 132, 137
Darwin, Charles, 5, 172
Datil National Forest: and deer, 70, 76, 115, 115–16n
Dauerwald, 141–42
Deer: adaptability of, 255–56; and agriculture, 124, 157, 252; browse, nutritive quality of, 143, 179, 180n; browse, palatability of, 158, 182, 253, 255; and cattle, 58, 69, 88–91, 113–15, 117, 120–21, 230n; damage to forests, 140, 142, 159, 160, 196, 233, 236–37, 253–54, 267; feeding, artificial, 142; feeding, winter, 157, 181, 192, 196; health, criteria of, 158, 177; overbrowsing by, 58, 81–82, 84–85, 89–91, 103, 106–8, 113–15, 140, 144–45, 147–50, 174, 179, 180–81, 184, 187, 189, 224, 245, 247, 253–54; overbrowsing by, in summer, 159, 263–64; sex ratio of, 80, 92–93, 97, 143, 158, 179, 185, 191, 192; social tolerance of, 251; and soil quality, 124, 127, 253; species identified, 39, 52, 69, 73, 123, 251; starvation of, 85, 144–45, 159, 174, 185, 186n, 187–91, 242, 257, 263; as symbol, 3, 53; too many (excess), 148–51, 171–72, 205, 262, 265; yarding areas, 124, 174, 187–90, 196, 255. *See also* Apache, Chequamegon, Gila, Kaibab, and other national forests; Chihuahua; Germany; "Huron Mountain Club, Report on"; Michigan; Pennsylvania deer; Rockford deer area;

Wisconsin deer; Coyotes, Environmental Change; Hunting; Irruptions; Kill factor; Predation; Productivity; Refuges; Stocking; Wolves
"Deer and Dauerwald in Germany," 139–44, 150
Deer Damage to Forest Reproduction Survey, 236–37
Deer habitat management, 125, 134, 147n, 158–59, 237–38, 244–45, 264–65
Deer management and forestry, 66–70, 88–89, 103–4, 125, 128–29, 134, 139–44, 161, 185, 244–45, 255, 264–65, 266–67. *See also* Wildlife management
"Deer Management in the Southwest," 71–75, 76, 87, 92, 94–95, 97
Deer research, 69–75, 92–93, 103–4, 118–21, 125, 130, 137, 139n, 154–56, 158–59, 173–74, 216–17, 243, 256. *See also* Wisconsin deer, research project
Deerwester, Richard, quoted, 196
Diamond Bar Ranch (range), 77, 79, 81, 90, 95, 96–98, 101, 111n
Dickinsen, Virgil, 188, 198, 230–31
Diversity, ecological, 34, 55n, 68, 80, 151; in Chihuahua, 153–54; concept of, 50–52; in Germany, 142–44; in Southwest, 36–37, 43, 50–52; and stability, 31–32, 164–67; in Wisconsin, 253, 264, 266, 267

Ecological analysis, difficulty of, 51–53, 74, 216–17, 256
Ecological attitude, 1–2, 17–18, 35, 151–52, 165–67, 172, 207–8, 216, 265–66, 268–71; toward predators, 95–96, 151
Ecological engineering, 169–72, 226
Ecological Society of America, 34
Ecology: history of, 5–7, 17, 24, 30, 151–52; and evolution, 30–31
Education, 168, 169, 172, 192, 206
Egler, Frank E., 7n
Elk, 38–39, 57–58, 62, 63–64
Elton, Charles, 24, 31n, 32n

Index

Index